Transistor Fundamentals

Volume 4
Digital and Special Circuits

by

LOUIS SCHWEITZER
and
REGINALD H. PENISTON

Under the direction of
Training & Retraining, Inc.

HOWARD W. SAMS & CO., INC.
THE BOBBS-MERRILL CO., INC.
INDIANAPOLIS · KANSAS CITY · NEW YORK

FIRST EDITION

THIRD PRINTING—1970

Copyright © 1968 by Howard W. Sams & Co., Inc., Indianapolis, Indiana 46206. Printed in the United States of America.

All rights reserved. Reproduction or use, without express permission, of editorial or pictorial content, in any manner, is prohibited. No patent liability is assumed with respect to the use of the information contained herein.

Library of Congress Catalog Card Number: 68:21313

Preface

One of the fastest growing and most widely used areas of electronics is that of transistorized digital circuits. Unlike the common amplifier, oscillator, and power circuits, the digital circuits find vast application in electronic computers, advanced direct readout test equipment, control circuits, and many other automated industrial processes.

This volume presents the fundamentals of the binary number system which is the language most commonly employed by digital circuits. It also presents a comprehensive study of frequency-generating circuits such as multivibrators, trigger circuits, and flip flops. Different types of AND, OR, NOT, inverter circuits and combinations of these circuits are explained in detail. In addition, the operating characteristics of some of the newer special-purpose developments in semiconductor devices such as zener diodes, photosensitive components, silicon controlled rectifiers, unijunction transistors, tunnel diodes and PNPN devices are explained.

You will find at the end of each topic and chapter a set of test questions and answers. They are designed to help you learn by reinforcing the facts presented in the text. A final examination has been included to test your comprehension after you have completed the book. It will also serve as a valuable review.

TRAINING & RETRAINING, INC.

Acknowledgments

Grateful acknowledgment is made to all those who participated in the preparation, compilation, and editing of this series. Without their valuable contributions this series would not have been possible.

Credit for the initial concept of the programmed learning techniques goes to Stanley B. Schiffman, staff member of Training & Retraining, Inc.

Finally, special thanks are due to the publisher's editorial staff for invaluable assistance beyond the normal publisher-author relationship.

                        SEYMOUR D. USLAN, Editor-in-Chief
                        and
                        HERMAN SCHIFFMAN, President
                        Training & Retraining, Inc.

ABOUT THE AUTHORS

Louis Schweitzer was born in Yugoslavia in 1920 and attended high school in New York City. He joined the U.S. Army in 1943 and was awarded the Bronze Star medal for military heroism. Prior to his discharge in 1946 he completed several radio electronic courses. Mr. Schweitzer has also completed several industrial electronics courses at ITT Federal Laboratories, Training and Education Center. He attended New York State University, Farmingdale. Currently, Mr. Schweitzer is a technical writer for the U.S. Army at Ft. Monmouth, New Jersey. He has written technical manuals for the U.S. Army Electronics Command and has been an instructor at the U.S. Army Signal School.

Reginald H. Peniston is Supervisory Equipment Specialist with the Transmission Branch, Communications and ADP Division Maintenance Engineering Directorate, U.S. Army Electronics Command, Ft. Monmouth, New Jersey. Mr. Peniston began his electronics career repairing radio and radar equipment in the U.S. Army. He is a Director of the Electronic Institute licensed under the New Jersey Department of Education and is an instructor in vocational electronics. He has written technical manuals for the U.S. Army Electronics Command and has been an instructor at the U.S. Army Signal School.

Introduction

This volume describes the principles of operation of transistorized digital circuits. The digital circuit language or binary number system is described so that the application of this number system in digital circuits can be understood.

The text is intended to give a general understanding of the various digital circuits, so that as new more refined circuits are developed, the same principles can be applied. Once an understanding of these circuits has been obtained, their application in computers, test equipment, and tracking and sensing equipment can be easily understood.

The application of transistorized digital circuits is still in its infancy. As more and more uses for these very versatile circuits are found, a sound understanding of the basic operating principles will make their application much easier.

WHAT YOU SHOULD KNOW BEFORE YOU START

An understanding of semiconductor and transistor operation and circuit applications is helpful but not required. This volume is written so that students who have a general electronics experience or educational background can follow the subject matter easily. New terms are defined and explained as they are introduced. The only math requirement is that needed to understand the meaning of simple equations and the ability to manipulate these equations.

WHY THE PROGRAMMED TEXT FORMAT WAS CHOSEN

During the past few years, new concepts of learning have been developed under the common heading of programmed instruction. Although there are arguments for and against each of the several formats or styles of programmed textbooks, the value of programmed instruction itself has been proved to be sound. Most educators now seem to agree that the style of programming should be developed to fit the needs of teaching the particular subject. To help you progress successfully through this volume, a brief explanation of the programmed format follows.

Each chapter is divided into small bits of information presented in a sequence that has proved best for learning purposes. Some of the information bits are very short—a single sentence in some cases. Others may include several paragraphs. The length of each presentation is determined by the nature of the concept being explained and by the knowledge the reader has gained up to that point.

The text is designed around two-page segments. Facing pages include information on one or more concepts, complete with illustrations designed to clarify the word descriptions used. Self-testing questions are included at the end of each of these two-page segments. These questions are in the form of statements requiring that you fill in one or more missing words. Answers are given at the top of the succeeding page, so you will have the opportunity to check the accuracy of your response and verify what you have or have not learned before proceeding. When you find that your answer to a question does not agree with that given, you should restudy the information to determine why your answer was incorrect. As you can see, this method of question-answer programming ensures that you will advance through the text as quickly as you are able to absorb what has been presented.

HOW YOU SHOULD STUDY THIS TEXT

Naturally, good study habits are important. You should set aside a specific time each day to study, in an area where you can concentrate without being disturbed. Select a time when you are at your mental peak, a period when you feel most alert.

Here are a few pointers you will find helpful in getting the most out of this volume.

1. Read each sentence carefully and deliberately. There are no unnecessary words or phrases; each sentence presents or supports a thought which is important to your understanding of electricity and electronics.
2. When you are referred to or come to an illustration, stop at the end of the sentence you are reading and study the illustration. Make sure you have a mental picture of its general content. Then continue reading, returning to the illustration each time a detailed examination is required. The drawings were especially planned to reinforce your understanding of the subject.
3. At the bottom of the right-hand pages you will find one or more questions to be answered. These contain "fill-in" blanks. In answering the questions, it is important that you actually do so in writing, either in the book or on a separate sheet of paper. The physical act of writing the answers provides greater retention than merely thinking the answer. Writing will not become a chore since all answers are short.
4. Answer all questions in a section before turning the page to check the accuracy of your responses. Refer to any of the material you have read if you need help. If you do not know the answer, even after a quick review of the related text, finish answering any remaining questions. If the answers to any questions you skipped still have not come to you, turn the page and check the answer section.
5. When you have answered a question incorrectly, return to the appropriate paragraph or page and restudy the material. Knowing the correct answer to a question is less important than understanding why it is correct. Each section of new material is based on previously presented information. If there is a weak link in this chain, the later material will be more difficult to understand.
6. Carefully study the Summary Questions at the end of

each chapter. This review will help you gauge your knowledge of the information in the chapter and actually reinforce your knowledge. When you run across questions you do not completely understand, reread the sections relating to these statements, and recheck the questions and answers before going to the next chapter.
7. Complete the final test at the end of the book. This test reviews the complete text and will offer you a chance to find out just what you have learned. It also permits you to discover your weaknesses and initiate your own review of the volume.

This volume has been carefully planned to make the learning process as easy as possible. Naturally, a certain amount of effort on your part is required if you are to obtain maximum benefit from the book. However, if you follow the pointers just given, your efforts will be well rewarded, and you will find that your study will be a pleasant and interesting experience.

Contents

CHAPTER 1

DIGITAL CIRCUITS 13

    Ancestors of Digital Circuits 14
    Digital Arithmetic 16
    Solid-State Logic—Circuit Applications 18
    Basic Elements of Digital Computers 20
    Number Systems 22
    Decimal-to-Binary Conversion 28
    Binary-to-Decimal Conversion 30
    Binary Addition 32
    Binary Subtraction 34
    Complements 36
    Binary-Coded Decimal Numbers 38
    Representation of Binary Numbers 40
    Multivibrators 42
    Counters 44
    Logic Circuits 46
    Symbolic Logic 48
    Basic Symbolic Logic Operations 50
    Combinational Circuits 54
    Special Semiconductor Circuits 56
    Summary 58

CHAPTER 2

LOGIC CIRCUITS 67

    Semiconductor Diodes and Biasing 68
    Diode Logic 70
    Transistor Types and Configurations 72
    Transistor Biasing 74
    Transistor AND Gates 76
    Transistor OR Gates 78
    Transistor NAND Gate 80
    Transistor NOR Gate 82
    Transistor Exclusive OR Circuit 84
    Combinational Circuits 86
    Duality of Logic Circuits 88
    Logic Diagrams and Equations 90
    Truth Tables and Logic 92
    Summary 94

CHAPTER 3

MULTIVIBRATOR CIRCUITS 103

    Why the Multivibrator is Required 104
    Collector-Coupled Multivibrator Circuit Operation . . 108
    Collector-Coupled Multivibrator Waveform Analysis . . 112
    Monostable (One-Shot) Multivibrator 114
    One-Shot Waveform Analysis 118
    Eccles-Jordan Multivibrator 120
    Eccles-Jordan Circuit Operation 122
    Logical Flip-Flop 124
    Sequential Circuits 126
    Summary 128

CHAPTER 4

WAVESHAPERS AND COUNTERS 135

    Pulse Generation 136
    Amplitude-Limiting Circuits 138
    Differentiator and Integrator Circuits 140
    Cascaded Binary Circuits 142
    Ring Counters 144
    Divide-By-Four Parallel Counter 146
    Divide-By-Three Parallel Counter 148
    Divide-By-Five Parallel Counter 150
    Parallel Decade Counter 152
    Counter Applications 154
    Summary 156

CHAPTER 5

SPECIAL SEMICONDUCTOR DEVICES 165

    Zener Diodes 166
    Zener Diode Applications 168
    Special Purpose Semiconductors 172
    Silicon Controlled Rectifier 176
    Unijunction Transistor 178
    UJT Circuit Applications 180
    Tunnel Diode Characteristics 182
    Tunnel Diode AND Gate and OR Gate 184
    PNPN Devices 186
    Summary 190

FINAL TEST 198

INDEX 205

1
Digital Circuits

What You Will Learn
In this volume you will learn about semiconductor circuits used in industrial control systems, monitor and switching networks in communications, and digital computers. Solid-state circuitry enables computers in conjunction with communication networks to control orbiting satellites, as shown below. Fig. 1-1 is intended to suggest the potentials of semiconductor applications, which are limited only by man's imagination. The principles of solid-state circuit operation that are applicable to logic circuits will be emphasized. The concepts presented will help you to understand present and, hopefully, future semiconductor circuits.

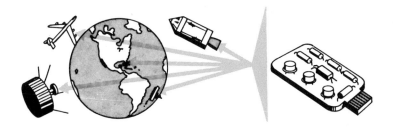

Fig. 1-1. Semiconductors in communications and control.

ANCESTORS OF DIGITAL CIRCUITS

"Digit" is derived from the Latin word *digitus* which means a finger or toe. The use of fingers (and presumably toes) for counting resulted in digits meaning the symbols 1 through 9 with 0 as the reference digit. *Digital circuits* perform arithmetic calculations with numbers. The use of semiconductors in digital circuits stems from our need for compact equipment that performs efficiently, reliably, and quickly. Electronic equipment development for purposes of digital calculations is based on the mechanical digital devices shown in Fig. 1-2.

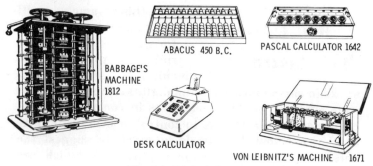

Fig. 1-2. Digital devices.

Development of Digital Devices

The abacus was developed about 2500 years ago to aid persons dealing with calculations. Refinements of these marvelous devices are used today. Surely, the abacus was a significant development in a series of digital devices.

In 1642, Blaise Pascal, the brilliant French philosopher and scientist, built the first mechanical adding machine. Pascal's machine consisted of wheels arranged so that turning one notched wheel ten notches (units) caused a second wheel to advance one notch (tens). Pascal's machine could add or subtract, depending on which direction the wheels were turned. Multiplication was performed by repeated addition; division by repeated subtraction. The same principle is used today in automobile mileage meters (odometers), cash registers, and other mechanical counting devices.

In 1672, the eminent German philosopher and mathematician, Gottfried Wilhelm von Leibnitz, invented a calculat-

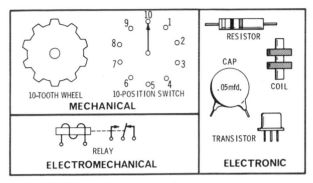

Fig. 1-3. Development of digital devices.

ing machine that directly performed calculations in multiplication, division, and the extraction of square roots. Von Leibnitz's "stepped reckoner" was the forerunner of the modern desk calculating machines.

In the early 1800's, Charles Babbage devised plans for a machine capable of automatically following instructions and solving arithmetic problems. The English inventor's machine did not become operational mostly because fabrication techniques of his time were not adequate. However, many of the important principles of today's digital computers are based on the work of a nineteenth century inventor.

Mechanical to Electromechanical to Electronic

Digital devices rapidly developed from bulky mechanical assemblies to less cumbersome, faster, semiautomatic units operated by electromechanical (relay) devices. Strictly mechanical or electromechanical digital devices are still used widely. However, when applications require almost instantaneous control or calculations, electronic components are used in place of the mechanical and electromechanical parts. Fig. 1-3 illustrates the transition from mechanical to solid-state devices.

Q1-1. Digital devices perform _____ with numbers.

Q1-2. When almost instantaneous results are required, _____ components are used rather than mechanical or electromechanical parts.

Q1-3. The Latin word _____ means a finger or toe.

Your Answers Should Be:

A1-1. Digital devices perform **calculations** with numbers.

A1-2. When almost instantaneous results are required, **electronic** components are used rather than mechanical or electromechanical parts.

A1-3. The Latin word **digitus** means a finger or toe.

DIGITAL ARITHMETIC

An understanding of the decimal and binary number systems is necessary to comprehend the operation of transistors in control and counting circuits. All of us can work with the digits 0 through 9 and perform addition, subtraction, multiplication, and division with these digits. However, electronic circuits are necessarily designed to handle digits differently than humans. For example, you mentally multiply 7 by 6 to obtain a product of 42; electronic circuits count 7 + 7 + 7 + 7 + 7 + 7 to provide a sum of 42.

Number Systems

Number systems can be classified into two types: additive and positional. Roman numbers or symbols are examples of the purely additive type. The Roman symbols LXVII means L + X + V + I + I, which equals 67 in the decimal system.

The decimal-number system is a combination of the additive and positional types. For example, the 1 in 15 has a different value than the 1 in 51. In 15, the 1 because of its position is actually ten-ones; the 1 in 51 because of its position is one-one or 1. Some other examples are shown in Fig. 1-4.

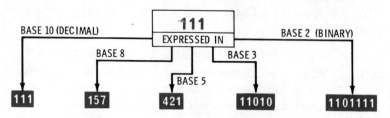

Fig. 1-4. Representation of number systems.

Need for Binary Number System

Either of two conditions, such as on or off, go or no-go (stop), high or low, yes or no, can be represented by the binary number system. Transistors can be operated as on-off switches which permit or prevent current. A transistor can be on (saturated) or off (cut-off). The condition of a switching circuit, consisting of many diodes, transistors, and associated electronic components, is often expressed in the binary number system.

The *base* or *radix* of any number system is determined by the number of different digits used. In the base 10 (decimal) number system, the ten digits 0 through 9 are used. In the base 2 (binary) number system, the two digits 0 and 1 are used.

Logic circuits, also called *switching circuits*, require an understanding of the binary number system. An outstanding characteristic of logic or switching circuits is that these circuits are composed of elements which behave like simple on-off switches.

Figs. 1-4 and 1-5 indicate the necessity for knowing the base of the number system being used in order to understand the quantity expressed by digits.

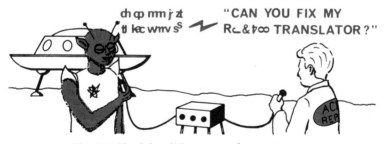

Fig. 1-5. Need for different number systems.

Q1-4. The Roman number system is of the _____ type; the decimal and binary number systems are of the _____ type.

Q1-5. The _____ of any positional-number system determines the number of digits used.

Q1-6. Elements that behave like simple on-off switches are used in _____ or _____ circuit.

> **Your Answers Should Be:**
> **A1-4.** The Roman number system is of the **additive** type; the decimal and binary number systems are of the **positional** type.
> **A1-5.** The **base** of any positional-number system determines the number of digits used.
> **A1-6.** Elements that behave like simple on-off switches are used in **logic** or **switching** circuits.

SOLID-STATE LOGIC CIRCUIT APPLICATIONS

The widespread application of logic circuits to business, industrial, scientific, and government activities is phenomenal. Although the uses for semiautomatic and automatic devices are increasing, so is the problem of grouping the applications. For convenience, two groups of applications will be discussed. The first group includes logic circuits in Automatic Data Processing (ADP) equipment and in digital computers. The second logic circuit applications group is called *real-time control*.

Automation

ADP equipment performs calculations for paychecks, inventories, invoices, and other repetitious and ordinarily time-consuming tasks. Before ADP, adequate and satisfactory accounting methods existed and are still being used by smaller businesses. The point is that ADP represents the trend toward using electronic equipment to do the job formerly performed by mechanical means. See Fig. 1-6.

Fig. 1-6. ADP and computers.

Digital computers automatically solve a series of arithmetic operations in contrast to devices that solve only a single arithmetic operation as shown previously. ADP is to accounting what a digital computer is to solving scientific problems. For example, calculating the orbit of a satellite requires the solution of thousands of individual arithmetic problems. Any person with a grade-school education can perform the arithmetic. However, he would be quite old before providing the final answer. Our efforts to probe space would be doomed without the use of digital computers.

Real-Time Control

A system which receives information (data) and automatically provides an answer in time to do something about the problem is a *real-time control system*. Fig. 1-7 below shows that certain manufacturing processes require real-time control of physical activities.

Perhaps you have noticed the overall size and apparent complexity of ADP equipment, computers, and control systems. Consider, however, the large number of logic and associated circuits required to perform the necessary functions. The equipment contains many components but there is much repetition involving only a few basic logic circuits.

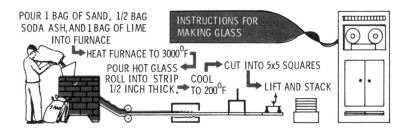

Fig. 1-7. Control of manufacturing processes.

Q1-7. ADP means _____ _____ _____.

Q1-8. ADP equipment is to accounting what a digital computer is to _____ scientific problems.

Q1-9. A system which automatically provides an answer in time to do something about the problem is a (an) _____-_____ _____ system.

> Your Answers Should Be:
> **A1-7.** ADP means **Automatic Data Processing**.
> **A1-8.** ADP Equipment is to accounting what a digital computer is to **solving** scientific problems.
> **A1-9.** A system which automatically provides an answer in time to do something about the problem is a **real-time control** system.

BASIC ELEMENTS OF DIGITAL COMPUTERS

The terminology describing the functioning of computers is common to the expressions used in logic circuits for other applications. Fig. 1-8 illustrates the basic functional elements of digital computers. Most present-day computers vary in circuit construction but the overall function remains the same. Before a computer can be placed in operation, a person called a *programmer* must translate or interpret the mathematical problem into terms the computer can understand. The language used by a computer is called its *code*. The programmer must provide the computer with data words, instruction words, and addresses. The *data words* represent information to be operated on; *instruction words* represent the operations to be performed. *Addresses* identify the precise storage locations of the data words and instruction words in the memory elements of the computer. A computer which stores its own control information is known as a *stored program* computer. Practically all modern digital computers are of this type. Exceptions can be found in the smaller lower-priced, special-purpose computers.

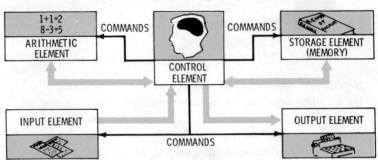

Fig. 1-8. Elements of digital computers.

Input Element

This functional element is a communication link between the programmer and the control element. Punched cards, punched tape, and magnetic tape are typical devices used with the input element.

Control Element

The control element is the brain of the computer. It decodes the input program and generates signals which tell the other functional elements what to do. The control element is unique because it is functionally part of the other elements.

Storage or Memory Element

This functional element is a "file cabinet" for data and instruction words. When the control element instructs the memory element to furnish information, the location within the memory element is identified by its address. Magnetic cores, tapes, and drums are the most common memory devices.

Arithmetic Element

This element performs the calculating function for the computer. The four basic arithmetic operations of addition, subtraction, multiplication, and division are performed by only two operations: addition and subtraction.

Output Element

This functional element is a communication link between the computer and the outside world. The output element translates the computer code to the numbers or letters intelligible to the user. Printing machines, magnetic tapes, and oscilloscopes are typical output devices.

Q1-10. The language of a computer is called its _____.

Q1-11. Information to be operated on is represented by _____ words.

Q1-12. Signals that tell the other functional elements what to do are generated by the _____ element.

> **Your Answers Should Be:**
> **A1-10.** The language of a computer is called its **code**.
> **A1-11.** Information to be operated on is represented by **data** words.
> **A1-12.** Signals that tell the other functional elements what to do are generated by the **control** element.

NUMBER SYSTEMS

The recording of quantity was one of man's most significant accomplishments. The history of the development of number systems relates man's efforts to overcome the difficulty of recording quantity and to simplify the rules for manipulating numbers arithmetically.

Since the hands are the most convenient tool nature has provided man, it was only natural that he used his fingers for counting. This was the beginning of decimal counting.

As shown in Fig. 1-9, using a single vertical mark to represent each unit counted was probably the first known method of systematic counting. This is known as the *tally* system; but it was too bulky for representing large numbers. The system was improved by making four vertical lines with a fifth diagonal line to form a set of five units.

The greatest advancement in the development of the decimal system came about when the Arabic symbols 0,1,2,3,4,5, 6,7,8, and 9 were adopted to represent the quantities zero to nine. When written in the order shown above, each Arabic symbol has a value of one more than the symbol preceding it.

These symbols, and the idea of giving each another value depending on its position, form the basis of our decimal system.

Fig. 1-9. Counting with tally.

Binary Digits

Fig. 1-10. Derivation of bits.

There are many number systems in use today in addition to the decimal system; however the system which is best suited to logic circuits is the *binary* system, a system using only two digits, 0 and 1. These two digits are commonly called bits which, as shown in Figure 1-10, is a contraction of the word "binary digits."

The binary system is used because so many of the circuits and components used in computers, control systems, etc., have just two conditions which are used to represent the two digits of the binary system. For example, a switch is either open, representing a 0, or closed, representing a 1. A relay is de-energized (0) or energized (1). A transistor, as illustrated in Fig. 1-11, is nonconducting, or represented by a 0; and when conducting is represented by a 1.

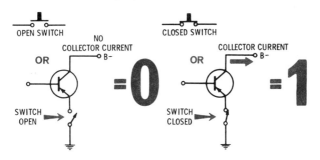

Fig. 1-11. Transistor switch, open or closed.

Q1-13. The ten Arabic symbols are used to represent quantities in the _____ number system.

Q1-14. The two digits 0 and 1 are used to represent quantities in the _____ number system.

Q1-15. A conducting transistor circuit represents a binary _____; a nonconducting circuit represents a binary _____.

> **Your Answers Should Be:**
>
> **A1-13.** The ten Arabic symbols are used to represent quantities in the **decimal** number system.
>
> **A1-14.** The two digits 0 to 1 are used to represent quantities in the **binary** number system.
>
> **A1-15.** A conducting transistor circuit represents a binary **1**; a nonconducting circuit represents a binary **0**.

Decimal System

The value of any digit is dependent on its base, or radix, and the position of the digit.

Positional Notation System

The positional notation of a number means that the value of each digit is determined by its position in the overall number, as seen in Fig. 1-12.

POSITION

MILLIONS	HUNDRED THOUSANDS	TEN THOUSANDS	THOUSANDS	HUNDREDS	TENS	UNITS
1 000 000	100 000	10 000	1 000	100	10	1

VALUE

Fig. 1-12. Decimal positional notation.

As an example, let's multiply 16 × 11. The actual meaning of the number 176 can be better understood if we see how it is spoken; "one-hundred and seventy-six." The number is actually (1 × 100) + (7 × 10) + 6. Here you can see that the value of each digit is determined by its position.

The positional values most commonly used are units to millions. Different words have been developed for the numbers 10 to 20; we say ten, eleven, twelve, thirteen, etc. From 20 on, the verbal representation of the numbers breaks only at the different positions (hundreds, thousands, millions, etc.). Fig. 1-13 shows what the notation 3333 really means.

THREE THOUSAND, THREE HUNDRED, AND THIRTY-THREE

Fig. 1-13. Value and position.

Although 3 is the only digit used, we can see that its position determines its value. It is necessary therefore to have only the ten basic digits and the positional notation system in order to count to any desired amount.

Using the information in the columns below, let us determine the positional values in the number 23,436. First, write down the numbers and the positional value of each number, then multiply the number by the positional value and add the results.

Positional Values					Multiply Number		Positional Value	
2	3	4	3	6	2	×	10,000 =	20,000
					3	×	1,000 =	3,000
Ten Thousands	Thousands	Hundreds	Tens	Units	4	×	100 =	400
					3	×	10 =	30
					6	×	1 =	6
								23,436

Base or Radix

The base or radix of a number system is the number of different digits that can be used in each position in the number system. The decimal system has a radix of ten; that is, there are ten different digits (0, 1, 2, 3, . . . 8, 9), any of which may be used in each position of the number. There are many number systems in use today, each having a different base or radix. The duodecimal system, having a radix of 12, is used for clocks, inches, and dozens. The binary system uses only two digits, a 0 or 1, in any digit position. Therefore two is the radix.

Q1-16. The values of each digit within a number is determined by its position in the overall number. This notation is called _____ notation.

Q1-17. The radix of the decimal system is _____.

Q1-18. The radix of the binary system is _____.

> **Your Answers Should Be:**
>
> **A1-16.** The value of each digit within a number is determined by its position in the overall number. This notation is called **positional** notation.
>
> **A1-17.** The radix of the decimal system is **ten**.
>
> **A1-18.** The radix of the binary system is **two**.

The Binary System

The binary system has several things in common with the decimal system; it has a radix and it also uses the same type of positional notation system.

Base or Radix—The base or radix of the binary system, as previously mentioned, is two. That is, there are only two different digits that can be used in each position of the binary number system. These two digits are 0 and 1.

Positional Notation—Just as in the decimal system, the value of each digit is determined by its position in the overall number. A review of the decimal system and the next drawing (Fig. 1-14) show that in the decimal system each position increases by ten over the previous position, while in the binary system each position increases by two.

As an example, the decimal number 125 is actually $(1 \times 100) + (2 \times 10) + 5$. This same number, 125, is represented in the binary system as 1111101. Referring to Fig. 1-14, the binary number 1111101 can be represented as $(1 \times 64) + (1 \times 32) + (1 \times 16) + (1 \times 8) + (1 \times 4) + (0 \times 2) + (1 \times 1) = 125$.

This can be simplified by the use of the binary system and the digits 0 and 1. If a 1 appears in the digit position,

POSITION

SIXTY-FOUR	THIRTY-TWO	SIXTEEN	EIGHT	FOUR	TWO	UNITS
64	32	16	8	4	2	1

VALUE

Fig. 1-14. Binary positional notation.

the value of that digit position is added to the value of all other digit positions containing a 1. If a digit position contains a 0, the value of that digit position is ignored. The binary number 1111101 can be represented as: one 64, plus one 32, plus one 16, plus one 8, plus one 4, plus one 1, equals 125.

Determine the positional values in the binary number 1010. First, write down the binary digits and the positional value of each digit, then multiply the digit by the positional value and add the results.

Positional Values				Multiply	Digit	×	Positional Value
1	0	1	0		1	×	8 = 8
					0	×	4 = 0
Eight's	Four's	Two's	Units		1	×	2 = 2
					0	×	1 = 0
							10

Fig. 1-15 below shows the binary equivalents of the decimal numbers 1 through 10.

Quantities expressed in any number system can be converted to any other number system, providing that the bases or radices of the number systems are known. Since we are concerned particularly with the binary- and decimal-number systems, short cuts to enable you to convert from one to the other will be explained.

DECIMAL	BINARY	DECIMAL	BINARY
1	1	6	110
2	10	7	111
3	11	8	1000
4	100	9	1001
5	101	10	1010

Fig. 1-15. Decimal-to-binary equivalents.

Q1-19. In the decimal system, each position increases by _____; in the binary system, each position increases by _____.

Q1-20. The only two digits used in the binary system are _____ and _____.

Q1-21. The binary number 1001, is the same as the decimal number _____.

27

> **Your Answers Should Be:**
>
> **A1-19.** In the decimal system, each position increases by **ten**; in the binary system, each position increases by **two**.
>
> **A1-20.** The only two digits used in the binary system are 0 and 1.
>
> **A1-21.** The binary number 1001 is the same as the decimal number **nine**.

DECIMAL-TO-BINARY CONVERSION

The conversion of decimal numbers to their binary equivalents can be remembered easily if these two facts are kept in mind:

First... the radix of the binary system is **two**.

Second... the digits 0 and 1 are the **only** digits used.

Let us convert the decimal number seven (7) to its binary equivalent to show how it's done. Referring to Fig. 1-16 below, you will see that we have a chart showing the positional values used in the positional notation of the binary system. The decimal number 7 contains 1 four, so a 1 is placed in the four position; 1 two, and 1 unit or one, so a 1 is placed in these two positions. Adding the positional values of the columns as a check, we see that binary 111 is the same as decimal 7.

Taking another example, to convert decimal 19 to a binary number, we see that decimal 19 contains 1 sixteen, no eights (16 plus 8 would equal more than 19), no fours (16 plus 4

POSITION

SIXTY-FOUR	THIRTY-TWO	SIXTEEN	EIGHT	FOUR	TWO	UNITS
64	32	16	8	4	2	1

VALUE

DECIMAL 7 = BINARY 111

Fig. 1-16. Decimal-to-binary conversion, limited method.

would equal more than 19), 1 two and 1 one. Therefore the binary 10011 is equivalent to decimal 19.

Conversion of decimal numbers to binary numbers by using a positional notation chart is fine for small decimal numbers, but as the decimal numbers get larger, the use of the table becomes a tedious and painstaking job. An easier way is to use the divide-by-two method. To use this method, divide the decimal number to be converted to binary by two, and the answer again by two, and so on, until you have a remainder of only one.

For example, refer to Fig. 1-17 and convert the decimal number 25 into its binary equivalent. To obtain the binary equivalent, read the remainders from BOTTOM TO TOP. Thus, the binary equivalent of 25 is 11001. Notice that throughout each step of division, all numbers are either exactly divisible by two or divisible by two with a remainder of 1. If two divides evenly into a number, place a 0 to the right of that division. When two does not divide evenly, place a 1 to the right of the division. Continue until further division by two is impossible. Read the answer (the binary number from BOTTOM TO TOP.

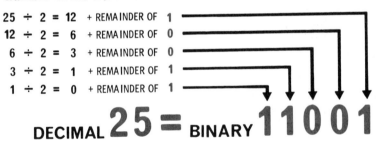

Fig. 1-17. Divide-by-two method.

Q1-22. A short-cut method for decimal-to-binary conversion of large decimal numbers is the _____-_____-_____ system.

Q1-23. In the divide-by-two method of decimal-to-binary conversion, the answer is read from _____ to _____.

Q1-24. The binary equivalents of the decimal numbers 27, 92, and 113 are _____, _____, and _____.

Your Answers Should Be:

A1-22. A short cut method for decimal to binary conversion of large decimal numbers is the **divide-by-two** method.

A1-23. In the divide-by-two method of decimal to binary conversion the answer is read from **bottom** to **top**.

A1-24. The binary equivalents of the decimal numbers 27, 92, and 113 are **11011, 1011100,** and **1110001**.

BINARY-TO-DECIMAL CONVERSION

Just as in the conversion of a decimal number to a binary number, a positional-notation chart can be used in the conversion of binary numbers to decimal numbers. First, write the digits of the number in a row, and then directly below them write the positional value of each digit of the binary number. Let us convert the binary number 101010 to its decimal equivalent to show how it is done.

Referring to Fig. 1-18, you will notice that in the conversion of 101010 to its decimal equivalent each digit of the binary number and its positional value has been recorded. We see that binary 101010 contains 1 thirty-two, 0 sixteens, 1 eight, 0 fours, 1 two, and 0 units or ones. Adding these values, we obtain the decimal number of 42.

Adding the positional values of a binary number to obtain its decimal equivalent can become too tedious and lengthy, especially in the case of large number. An easier way is to

POSITION

SIXTY-FOUR	THIRTY-TWO	SIXTEEN	EIGHT	FOUR	TWO	UNITS
64	**32**	**16**	**8**	**4**	**2**	**1**

VALUE

DECIMAL 32 + 0 + 8 + 0 + 2 + 0

42 = 1 0 1 0 1 0

Fig. 1-18. Binary-to-decimal conversion, limited method.

1 DOUBLE THE FIRST BIT, ADD THE DOUBLED VALUE TO THE NEXT BIT.

2 DOUBLE THE SUM OBTAINED, ADD THIS DOUBLED VALUE TO THE NEXT BIT.

3 CONTINUE STEP **2** UNTIL THE LAST BIT HAS BEEN ADDED TO THE PREVIOUSLY DOUBLED SUM.

EXAMPLE CONVERT BINARY 11001 TO DECIMAL 25

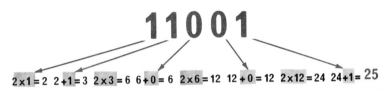

Fig. 1-19. Double-dadd method.

use the *double-dadd* method. Fig. 1-19 gives the rules for using the double-dadd method of binary-to-decimal conversion.

Using the rules given above let us convert binary 101010 to its decimal equivalent.

1. Double the first digit to get 2, add this doubled value to the next digit 0, to get 2.
2. Double the value obtained to get 4, add this value to the next digit 1, to get 5.
3. Double the value obtained to get 10, add this value to the next digit 0, to get 10.
4. Double the value obtained to get 20, add this value to the next digit 1, to get 21.
5. Double the value obtained to get 42, add this value to the next digit 0, to get 42.

The decimal equivalent of binary 101010 therefore is 42.

Q1-25. A short cut method to use for binary-to-decimal conversion of large binary numbers is the ____-_____ method.

Q1-26. Binary 1101101 is the same as decimal _____.

Q1-27. Binary 100011 is the same as decimal _____.

Q1-28. Binary 111111 is the same as decimal _____.

Q1-29. Binary 10011001 is the same as decimal _____.

> **Your Answers Should Be:**
>
> **A1-25.** A short cut method to use for binary-to-decimal conversion of large binary numbers is the **double-dadd** method.
>
> **A1-26.** Binary 1101101 is the same as decimal **109**.
>
> **A1-27.** Binary 100011 is the same as decimal **35**.
>
> **A1-28.** Binary 111111 is the same as decimal **63**.
>
> **A1-29.** Binary 10011001 is the same as decimal **153**.

BINARY ADDITION

Addition of binary numbers is the same as addition in the decimal system. The rules for addition of binary numbers are presented in Fig. 1-20.

RULES FOR BINARY ADDITION

RULE 1	0 + 0 = 0
RULE 2	1 + 0 = 1
RULE 3	0 + 1 = 1
RULE 4	1 + 1 = 0 WITH A CARRY OF 1

Fig. 1-20. Binary addition.

The first three rules are exactly the same as those for addition of decimal numbers and are easy to understand. Adding the binary number 11 to the binary number 100 will produce a binary sum of 111.

```
Binary System                    Decimal System
   100 ............................ 4
    11 ............................ 3
   ---                             ---
   111 ............................ 7
```

To better understand rule 4 let us first review how we "carry" a number in the decimal system. When two decimal numbers are added, for example 36 and 8, we start with the units column and total each column. Since 6 and 8 equal 14 (which is larger than 10) the 4 is brought down, and the 10 is carried to the tens column. The tens column is added (3 + 1) to obtain the total of 44.

```
 36
  8
 --
  4
 10 (carry)
 --
 44
```

32

```
    1 CARRY          1 CARRY          1 CARRY          1 CARRY
    1 0 0 1          1 0 0 1          1 0 0 1          1 0 0 1
    1 0 0 1          1 0 0 1          1 0 0 1          1 0 0 1
    ───────          ───────          ───────          ───────
          0              1 0            0 1 0          1 0 0 1 0
```

Fig. 1-21. Binary addition, carry rule.

The same principle of carrying also applies to binary addition. Rule 4 specifies that when you add 1 + 1 you obtain 10 (decimal 2), so you put down the 0 and carry the 1 to the next column to the left, as shown in Fig. 1-21.

There is another point that must be understood when applying rule 4. If, when carrying from one column to the next, the column to the left contains a 1, the limit for that column has been reached. Again a 0 is brought down with a carry of 1 to the next column to the left. Referring to Fig. 1-22, let us apply the rules for binary addition.

```
    1 CARRY         1 1 CARRY        1 1 CARRY        1 1 CARRY
    1 0 1 1          1 0 1 1          1 0 1 1          1 0 1 1
    1 0 0 1          1 0 0 1          1 0 0 1          1 0 0 1
    ───────          ───────          ───────          ───────
          0              0 0            1 0 0          1 0 1 0 0
```

Fig. 1-22. Application of addition rules.

Step A. Apply rule 4. Write 0 and carry 1 to the next column to the left.
Step B. Apply rule 4 again. Seeing that a 1 already appears in the next column to the left, change the 1 to a 0, and carry the 1 to the next column to the left. Then apply rule 1.
Step C. Substitute the 1 carried from the previous column for the 0, then apply rule 3.
Step D. Apply rule 4.

Q1-30. The four rules for binary addition are _____ + _____ = _____, _____ + _____ = _____, _____ + _____ = _____, and _____ + _____ = _____ with a carry of _____.

Q1-31. Add the following binary numbers:

    110011   101011   10000011
+101101   +111111   +11010001

> **Your Answers Should Be:**
> **A1-30.** The four rules for binary addition are: $0 + 0 = 0$, $0 + 1 = 1$, and $1 + 1 = 0$ with a carry of 1.
> **A1-31.** Add the following binary numbers:
>
> | | | |
> |---|---|---|
> | 110011 | 101011 | 10000011 |
> | +101101 | +111111 | +11010001 |
> | **1100000** | **1101010** | **101010100** |

RULES FOR BINARY SUBTRACTION

RULE 1   **0 − 0 = 0**
RULE 2   **1 − 0 = 1**
RULE 3   **1 − 1 = 0**
RULE 4   **0 − 1 = 1** WITH A BORROW OF **1**

Fig. 1-23. Binary subtraction.

BINARY SUBTRACTION

Fig. 1-23 gives the rules for subtraction of binary numbers.

The first three rules for binary subtraction are also easy to understand because they are exactly the same as those used for subtraction of decimal numbers. Subtracting the binary number 11 from the binary number 111 will produce a binary difference of 100. Check your subtraction by converting the binary numbers to their decimal equivalents using the double-dadd system.

Binary System	Decimal System
111	7
−11	−3
100	4

To understand rule 4, let us first review how we borrow in the decimal system. When two decimal numbers are subtracted, for example, 8 subtracted from 44 we start with the units column at the extreme right and subtract each column.

```
  0 1 BORROW      0 1           0 1           0 1
1 1 1̶ 0         1 1 1̶ 0       1 1 1̶ 0       1 1 1̶ 0
0 1 0 1         0 1 0 1       0 1 0 1       0 1 0 1
─────────       ─────────     ─────────     ─────────
       1            0 1         0 0 1       1 0 0 1
```

Fig. 1-24. Application of subtraction rules.

Since we cannot subtract 8 from 4, we must borrow ten from the tens column, making the units column 14. We can now subtract 8 from 14, to obtain an answer of 6. The 10 borrowed from the tens column now makes this column 3 instead of 4. Since there is nothing to subtract from the 3, the 3 is brought down, to obtain a total of 36.

```
  3 1 (borrow)
   44
    8
   ──
   36
```

The same principal of borrowing also applies to binary subtraction. Rule 4 specifies that when you subtract 1 from 0 you must first borrow a 1 from the next column to the left.

Referring to Fig. 1-24, let us apply the rules for binary subtraction:

Step A. Apply rule 4. We cannot subtract 1 from 0, so borrow 1 from the next column to the left. Put down a 1 in the answer, and change the 1 in the top row (next column to the left) to a 0.
Step B. Apply rule 1. 0 from 0 equals 0.
Step C. Apply rule 2. 1 from 1 equals 0.
Step D. Apply rule 3. 0 from 1 equal 1.

Check the answer by converting the binary numbers to decimal numbers.

Q1-32. The four rules for binary subtraction are _____ − _____ = _____, _____ − _____ = _____, _____ − _____ = _____, and _____ − _____ = _____ with a borrow of _____.

Q1-33. Subtract the following binary numbers:

```
  1001        10110        1100011
 −0111       −01011       −0111011
 ─────       ──────       ────────
```

35

> **Your Answers Should Be:**
> **A1-32.** The four rules for binary subtraction are
> $0 - 0 = 0$, $1 - 1 = 0$, $1 - 0 = 1$, and $0 - 1 = 1$
> with a borrow of **1**.
> **A1-33.** Subtract the following binary numbers:
>
> ```
>   1001        10110        1100011
>  -0111       -01011       -0111011
>   0010        01011        0101000
> ```

COMPLEMENTS

One of the most important factors in the design of equipment using logic circuits is to simplify the equipment as much as possible. Logic circuits can perform subtraction, but in order to accomplish this, elaborate circuits would be required. By using *complements,* subtraction can be reduced to addition. The two types of decimal complements are the 10's complement (true complement) and the 9's complement (radix-minus-one complement). The rules for both the true and radix-minus-one complements of the decimal number 325 are given below in Fig. 1-25.

TRUE:
A. SUBTRACT EACH DIGIT OF THE NUMBER FROM 9.
B. ADD 1 TO THE LEAST SIGNIFICANT DIGIT OF THE NUMBER OBTAINED

```
         999                674
STEP A. -325     STEP B. +    1
         674                675
                 10's COMPLEMENT OF 325
```

RADIX-MINUS ONE:
A. SUBTRACT EACH DIGIT OF THE NUMBER FROM 9.

```
         999
STEP A. -325
         674   9's COMPLEMENT OF 325
```

Fig. 1-25. Complement rules.

Subtraction Using True Complement Method

Subtraction using the true complement method is performed by following the two rules given next in Fig. 1-26.

1 ADD THE TRUE COMPLEMENT OF THE SUBTRAHEND TO THE MINUEND.
2 IF THERE IS A CARRY-OVER, ELIMINATE IT.

```
                168 MINUEND                        168 MINUEND
REGULAR        -051 SUBTRAHEND      TRUE          +949 10's COMPLEMENT OF 051
SUBTRACTION     117 DIFFERENCE      COMPLEMENT    ✗117 SUM
                                    METHOD
```

Fig. 1-26. Subtraction using true complement.

1 ADD THE RADIX-MINUS-ONE COMPLEMENT OF THE SUBTRAHEND TO THE MINUEND.

2 ADD THE CARRY-OVER TO THE LEAST SIGNIFICANT DIGIT.

```
                    168  MINUEND                           168  MINUEND
REGULAR            -051  SUBTRAHEND      RADIX-         + 948  9's COMPLEMENT OF 051
SUBTRACTION        ----                  MINUS-          ⓛ116  SUM
                    117  DIFFERENCE      ONE            +   ↲1  CARRY-OVER
                                         COMPLEMENT     -----
                                         METHOD          117  DIFFERENCE
```

Fig. 1-27. Subtraction using radix-minus-one complement.

Subtraction Using Radix-Minus-One Complement Method

Subtraction using the radix-minus-one complement method is performed by following the two rules given in Fig. 1-27.

```
1 0 1 0 1 1 1 0 1 = 0 1 0 1 0 0 0 1 0
      1 1 0 1 0 0 = 0 0 1 0 1 1
           1 1 1 = 0 0 0
```

Fig. 1-28. 1's complement of binary numbers.

Complement Method of Subtraction in the Binary System

Subtraction by complementing the subtrahend and adding can also be used in the binary system, except the 1's complement is used. The 1's complement of a binary number is easy to remember; simply replace all the 1's with 0's and all the 0's with 1's.

The next illustration (Fig. 1-29) shows the procedure for the complement method of subtraction in the binary system.

```
                 18   MINUEND                        10010   MINUEND
DECIMAL         -12   SUBTRAHEND      1's          + 0011    1's COMPLEMENT OF 1100
SUBTRACTION     ---                   COMPLEMENT     ⓛ0101   SUM
                  6   DIFFERENCE      METHOD       +   ↲1    CARRY-OVER
                                                    -----
                                                     110     DIFFERENCE
```

Fig. 1-29. Application of 1's complement.

Q1-34. The true complements of the following numbers are: 638 = _____, 891 = _____, 128 = _____.

Q1-35. The 9's complements of the following numbers are: 38 = _____, 386 = _____, 998 = _____.

Q1-36. Subtract the following numbers using the 1's complement method:

```
  110010        110111        11111
 -101100       -010100       -10101
```

Your Answers Should Be:

A1-34. The true complements of the following numbers are: 638 = **362**, 891 = **109**, 128 = **872**.

A1-35. The 9's complements of the following numbers are: 38 = **061**, 386 = **613**, 998 = **001**.

A1-36.
```
  110010        110111         11111
 +010011       +101011        +01010
 -------       -------        ------
 1000101       1100010        101001
       1             1             1
 -------       -------        ------
  000110        100011         01010
```

BINARY-CODED DECIMAL NUMBERS

Binary-coded representation of decimal numbers is a system in which each digit of a decimal number is represented by a combination of four *binary digits* or *bits*. Since it is not possible for logic circuits to manipulate decimal numbers directly, there are several binary-coded forms which are used. The two most commonly used are the *natural binary decimal code* and the *excess-3 code*.

Natural Binary Decimal Code

In this system each digit of a decimal number is represented by its binary equivalent as shown in Fig. 1-30.

Fig. 1-30. Natural binary decimal code.

Notice that this system differs from the pure binary system in that the decimal number is converted to its four-bit binary equivalent on a digit-by-digit basis. The decimal number 312 as shown in Fig. 1-30, expressed in pure binary, would be 100111000.

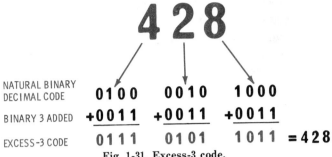
Fig. 1-31. Excess-3 code.

Excess-3 Code

In this system each digit of the decimal number is represented by its binary equivalent plus 3 (0011), as shown in Fig. 1-31.

To reconvert the excess-3 coded number back to its decimal equivalent, 3 (0011) must be subtracted from each four-bit binary group to obtain the natural binary decimal code. Then each four-bit binary group must be converted to its decimal equivalent (Fig. 1-32).

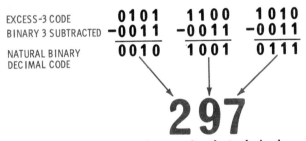
Fig. 1-32. Conversion of excess-3 code to decimal.

Q1-37. Represent the following numbers in natural binary decimal code: 21 = _____, 106 = _____, 268 = _____.

Q1-38. Represent the following numbers in excess-3 code: 16 = _____, 361 = _____, 58 = _____.

Q1-39. Convert the following excess-3 coded numbers their decimal equivalents:
0101-1001 = _____, 0100-0110-0100 = _____,
0111-1011-1000 = _____.

> **Your Answers Should Be:**
>
> **A1-37.** Represent the following numbers in the natural binary decimal code: 21 = **0010-0001**, 106 = **0001-0000-0110**, 268 = **0010-0110-1000**.
>
> **A1-38.** Represent the following numbers in excess-3 code: 16 = **0100-1001**, 361 = **0110-1001-0100**, 58 = **1000-1011**.
>
> **A1-39.** 0101-1001 = **26**, 0100-0110-0100 = **131**, 0111-1011-1000 = **485**.

REPRESENTATION OF BINARY NUMBERS

Throughout our discussion of binary numbers, we found that the digits 0 and 1 could be used to represent any binary quantity. This is fine for paper and pencil calculations, but electrical signals must be used to apply the desired information to transistor logic circuits.

There are two electrical signals that can be used to represent the binary bits 0 and 1. Because transistor circuits used in control devices and computers require speed and accuracy, the two signals must meet very rigid requirements. First, they must be suitable for use in high-speed circuitry. Second, the two signals must be very easy to tell apart. Third, they must be hard to confuse with each other.

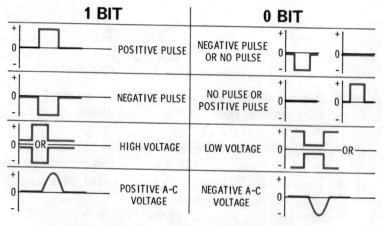

Fig. 1-33. Signals used to represent binary bits.

The second and third requirements may appear to be the same, but they are not. All transistor circuits, to some degree, distort signals that pass through them. Some signals can look confusingly alike after they have been distorted by many transistor circuits.

Fig. 1-33 shows several of the signal pairs that meet the three requirements stated above. As seen in this figure it is almost impossible to distort a positive pulse (representing a 1) to make it look like no pulse, or a negative pulse (representing a 0).

Depending on the type of electrical input used, a-c or d-c signals to represent the binary numbers, the two extreme conditions of the signal voltage are used. When a positive d-c voltage is used, the extremes are 0 and maximum positive. With negative d-c voltage the extremes are 0 and maximum negative. When an a-c signal is used, the extremes are maximum positive and maximum negative.

Fig. 1-34 shows several examples of signal pairs used to represent the bits of a binary number.

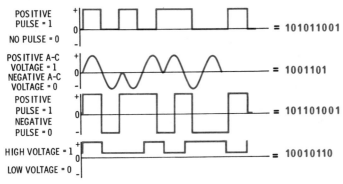

Fig. 1-34. Signal voltages representing binary numbers.

Q-40. Binary digits, when applied to transistor circuits, are represented by _____ _____.

Q1-41. When a maximum positive d-c voltage is used for binary 1, a _____ _____ or a _____ d-c voltage would be used for a binary 0.

Q1-42. When a negative voltage is used for a binary 0, a _____ _____ or a _____ voltage represents a 1.

> **Your Answers Should Be:**
>
> **A1-40.** Binary digits, when applied to transistor circuits, are represented by **electrical signals**.
>
> **A1-41.** When a maximum positive d-c voltage is used for a binary 1, a **less positive** or a **negative** d-c voltage would be used for a binary 0.
>
> **A1-42.** When a negative voltage is used for a binary 0, a **less negative** or a **positive** voltage represents a 1.

MULTIVIBRATORS

Multivibrators as used with logic circuits can be classified into two types, the *flip-flop* and the *monostable* multivibrators. Although it is a bistable or two-state device, the flip-flop differs from the basic logic circuits that will be described later in this chapter. Basically, the *flip-flop* is a storage device for bits. Monostable multivibrators include the *one-shot* and the *Schmitt trigger*. Essentially, monostable multivibrators generate time delays. The time that elapses between the input and the output of the monostable multivibrator is called the *delay time*.

Basic Flip-Flops

Fig. 1-35 shows the symbols for two kinds of flip-flops. A flip-flop can be in either one of two possible states; its output representing either a 1 or 0. If the output is in the 0 state, an S (set) input signal will set the flip-flop output to the 1 state. If the flip-flop output is in the 1 state an S input signal will not change the state of the output. When the output is in the 1 state, an R (reset) input signal will reset the flip-flop output to the 0 state. If the flip-flop output is in the 0 state, an R input signal will not change the state of the output.

A third input to the flip-flop in Fig. 1-35 is marked T, indicating a *trigger* input signal. A trigger input signal changes the state of the flip-flop. For example, if the flip-flop is in the 0 state, the T input signal changes the flip-flop output to the 1 state.

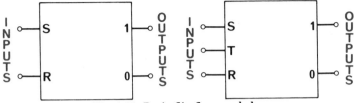

Fig. 1-35. Basic flip-flop symbols.

Monostable Multivibrators

The symbols for two kinds of monostable multivibrators are shown in Fig. 1-36. In contrast to the flip-flop, the one shot and the Schmitt trigger (ST) have only one input signal. They are similar to flip-flops in that all are two-state devices. The normal output state of either the one-shot or the Schmitt trigger with no input signal is the 0 state.

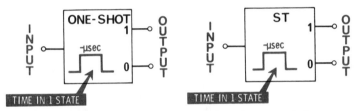

Fig. 1-36. One-shot and Schmitt-trigger symbols.

When an input signal is applied, the state of the output is changed from 0 to 1. The duration of time (time delay) that the one shot or ST remains in the 1 state is determined by internal circuit characteristics. That is, the input signal does not determine the time delay generated. The duration of time that the device remains in the 1 state is usually indicated on the symbol. Up to this point, the one shot and ST appear to be functionally identical. This is true, except for current-handling capabilities. Circuit operation and analysis of multivibrators will be explained later.

Q1-43. A flip-flop _____ bits; a monostable multivibrator generates _____ _____.

Q1-44. The letters S, T, and R respectively, mean ____, _____, and _____.

Q1-45. The one-shot and ST are similar to flip-flop because they are _____-_____ devices.

> **Your Answers Should Be:**
> **A1-43.** A flip-flop **stores** bits; a monostable multivibrator generates **time delays**.
> **A1-44.** The letters S, T, and R respectively, mean **set, trigger,** and **reset**.
> **A1-54.** The one-shot and ST are similar to flip-flop because they are **two-state** devices.

COUNTERS

Many scientific, industrial, military, and commercial applications of electronic equipment require that the equipment count discreet events and either record these events or produce some action after a predetermined number of events have occurred. The counter is the device which meets this requirement.

The *preset* counter is the type of counter that produces one output bit after receiving the number of input bits as determined by a front-panel control setting.

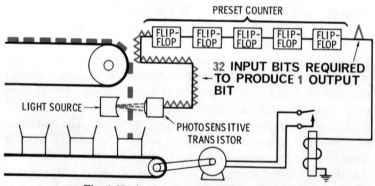

Fig. 1-37. Application of counter circuit.

In a machine-screw packaging operation as shown in Fig. 1-37, a preset counter is adjusted to produce one output bit after it receives 32 input bits. The input bits are produced as the machine screws fall from the conveyor belt, interrupting the beam from a light source to a photosensitive transistor. When the 32nd machine screw passes through the light beam the counter produces one output bit.

The main advantage of electronic counters over mechanical counters is that of speed. Another advantage is that the counting rate can be changed by simply setting controls. The mechanical counters require more extensive modifications. However, there are two types of electronic counters, the *binary* and the *decade counter*.

Binary Counters

The binary counter, as the name implies, is a counter which performs the function of counting in the binary system. The counter shown in Fig. 1-38 represents a five-stage binary counter that can count up to binary 11111 (decimal 31). The operation of binary and decade counters will be discussed in detail in Chapter 4.

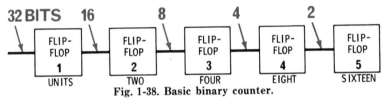

Fig. 1-38. Basic binary counter.

Decade Counters

A decade counter is a circuit having a counting ratio of ten; that is, the circuit produces one output pulse for every ten input pulses. The decade counter shown in Fig. 1-39, commonly called a *ring counter*, will also be discussed in detail in Chapter 4.

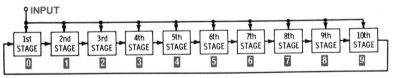

Fig. 1-39. Basic decade counter.

Q1-46. After receiving a predetermined number of input bits, a _____ _____ produces one output bit.

Q1-47. A six-stage binary counter can count up to binary _____ (decimal _____).

Q1-48. A decade counter produces one output pulse for every _____ input pulses.

> **Your Answers Should Be:**
> **A1-46.** After receiving a predetermined number of input bits, a **preset counter** produces one output bit.
> **A1-47.** A six-stage binary counter can count up to binary **111111** (decimal **63**).
> **A1-48.** A decade counter provides one output pulse for every **ten** input pulses.

LOGIC CIRCUITS

In logic or switching circuits, the information being processed is coded in terms of 1's and 0's. The 1's are represented by an electric potential of a certain value and the 0's by a potential of another value. Logic circuits operate on various bits of information, where a bit is a single electrical signal or some other representation which may be either 1 or 0 at any given time. The basic processes performed by logic circuits are AND, OR, and NOT.

AND Gate

The word *gate,* as used in this volume, means a circuit with one output and more than one input arranged so that the output is energized only when certain input conditions are met. As shown in Fig. 1-40, certain conventions must be established and explained. The lighted or extinguished condition of lamp DS1 (or any other load) can be expressed as either a 1 or a 0. In Fig. 1-40 we have assumed that a 1 represents lamp DS1 lighted (energized) and switches S1 and S2 as being closed (energized). An analysis of this simple series-connected circuit indicates that the lamp is energized (a 1 condition) only when both switches are closed. Also, the AND gate, or circuit, provides an output only when both input signals are present.

Fig. 1-40. AND gate symbol and equivalent circuit.

OR Gate

An OR gate or circuit consists of one output and two or more inputs. This type of circuit provides an output when any one of the input signals or voltage is present. Shown next in Fig. 1-41 is the symbol for an OR gate and the equivalent parallel switch-type circuit. An analysis of the simple parallel-connected circuit indicates that the lamp (load) is energized when switch $S1$ OR switch $S2$ is closed (energized).

Fig. 1-41. OR gate symbol and equivalent circuit.

NOT Circuit

The NOT circuit, also known as an inverter, consists of one output and one input. It provides no output when an input is present and produces an output when no input is present. In other words, a NOT circuit inverts or negates the input signal. Fig. 1-42 shows a symbol for the NOT circuit and the equivalent relay-type circuit. An analysis of the relay circuit indicates that the lamp (load) is energized when the coil of the relay is de-energized. That is, when the relay coil is not energized, the normally closed (NC) contact makes with the common (C) contact and the lamp (DS1) is energized.

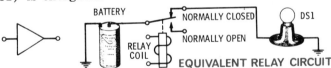

Fig. 1-42. NOT circuit symbol and equivalent circuit.

Q1-49. An AND gate is equivalent to a (an) _____ _____ switch-type circuit.

Q1-50. An OR gate is equivalent to a (an) _____- _____ switch-type circuit.

Q1-51. A NOT circuit _____ or negates the input signal.

47

> **Your Answers Should Be:**
> **A1-49.** An AND gate is equivalent to a **series-connected** switch-type circuit.
> **A1-50.** An OR gate is equivalent to a **parallel-connected** switch-type circuit.
> **A1-51.** A NOT circuit **inverts** or negates the input signal.

SYMBOLIC LOGIC

A brief historical background of symbolic logic will be presented, including an explanation of the need for this important and interesting field of study. Also, the relationship of symbolic logic to both logic circuits and the equivalent switch-type circuits will be explained.

Background and Need for Symbolic Logic

Our ability to communicate by words is limited by the loose meanings of these words. The problem of confusion in language has been recognized for thousands of years. Aristotle (384-322 BC) is credited with the development of the first formal logic system. He devised shorthand symbols and assumed all statements were either true or false. For example the letters MAP in Aristotelian logic mean "all men are mortal" since M means men, A means all are, and P means mortal.

In the 19th century, George Boole investigated the fundamental operations of the human mind and developed a language of symbols rather than words. Boole's investigation of the laws of thought forms the foundation for the field of study today known as symbolic logic, or Boolean algebra.

Logic Circuits and Symbolic Logic

The three basic logic circuits (AND, OR, and NOT) can be expressed as shown in Fig. 1-43.

$$F = A \text{ AND } B = A \times B = AB \qquad \text{NOT } 1 = 0$$

$$F = A \text{ OR } B = A + B \qquad \text{NOT } 0 = 1$$

Fig. 1-43. Symbolic logic expressions.

The symbolic-logic expression F equals AB means an AND gate or circuit provides an output (F) only when all input signals (A AND B) are present. The expression F equals A + B means an OR circuit provides an output (F) when one or more (A OR B) input signals are present. The expression NOT 1 means 0, which is provided at the output of a NOT (or inverter) circuit when a signal 1 is applied to the input. Also, the signal 1 is generated by a NOT circuit when the signal 0 is present at the input of the NOT circuit.

Equivalent Switch-Type Circuits and Symbolic Logic

Symbolic-logic expressions can be used to represent switch- and relay-type circuits that are equivalent to logic circuits. (Refer to Fig. 1-44). The symbols selected, A, B, C, and F, are strictly arbitrary. However, these symbols are two-valued such as A or A'. Each symbol can be only one of two possible values to represent one of two voltage levels. Since the binary number system is also two-valued, we can say that the symbol A represents a 1 and A' represents a 0.

Three basic processes are performed with the aid of symbolic logic: multiplication (AND); addition (OR); and negation or inversion (NOT), which is also called *complementation*. The process of addition with symbolic logic should not be confused with the addition of binary numbers. The processing of symbolic logic expressions will be explained.

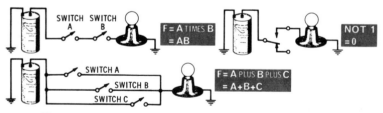

Fig. 1-44. Logic expressions for equivalent logic circuits.

Q1-52. When signal 0 (zero) is at the input of a NOT circuit, signal _____ is at the output.

Q1-53. When only signal B is at the input of a circuit expressed by F equals A + B + C, then _____ is at the output.

Q1-54. When signal X is present at the output of a NOT circuit, then signal _____ is at the input.

49

Your Answers Should Be:

A1-52. When signal 0 (zero) is at the input of a NOT circuit, signal **1** is at the output.

A1-53. When signal B is at the input of a circuit expressed by $F = A + B + C$, then **F** is at the output.

A1-54. When signal X is at the output of a NOT circuit, then signal **X'** is at the input.

BASIC SYMBOLIC LOGIC OPERATIONS

Multiplication, addition, and inversion (negation) are the operations or processes performed by AND, OR, and NOT circuits, respectively. Symbolic-logic notations or expressions are tools that permit us to represent word statements and logic circuits with symbols.

Logical Multiplication

As shown in Fig. 1-45, the expression $F = AB$ means that the AND gate provides an output (F) only when both inputs (A and B) are present at the same time. Since this AND gate has two inputs and each input can be either 1 or 0, there are four possible combinations of A and B. It is assumed that the presence of a pulse can be represented by a 1 and the absence of a pulse by a 0. Then, $A = 1$ or $A' = 0$ (zero); likewise, $B = 1$ or $B' = 0$; also, $F = 1$ or $F' = 0$. Fig. 1-45 lists all input combinations and the resultant outputs of the AND gate. All of the possible situations are listed below.

1. $A' \times B' = A'B' = B'A' = 0 \times 0 = 0 = F'$ (no output).
2. $A' \times B = A'B = BA' = 1 \times 0 = 0 \times 1 = 0 = F'$.
3. $A \times B' = AB' = B'A = 0 \times 1 = 1 \times 0 = 0 = F'$.
4. $A \times B = AB = BA = 1 \times 1 = 1 = F$ (output).

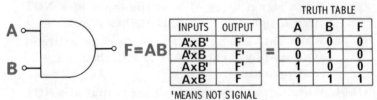

'MEANS NOT SIGNAL

Fig. 1-45. Logical multiplication.

Fig. 1-46 shows that 1 × A equals A and that 1 × A' equals A'. Although two inputs are shown in this illustration, the 1 input is not a variable. That is, the 1 input is not two-valued, but fixed at the 1 level. As shown in the switch-type circuit, the tabulation, and the timing diagram, an output from this circuit is entirely dependent on the presence (or absence) of signal A (or A'). An analysis of F equals A × 1 indicates this AND circuit can be replaced by a conductor with signal A at both ends, since F equals 1 × A equals A.

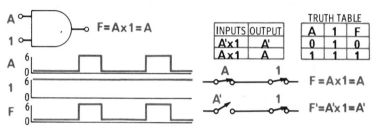

Fig. 1-46. Proof that 1 times A equals A.

Analyze the hypothetical AND circuit shown in Fig. 1-47. Although two inputs are shown, only signal A is two-valued. This AND circuit can be replaced by an open switch or some

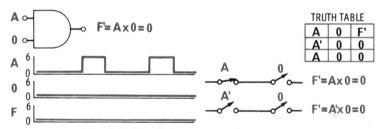

Fig. 1-47. Proof that 0 times A equals 0.

other device that will inhibit signal A from reaching the output. In this case, the output is entirely dependent on 0 since 0 × A or 0 × A' equals 0.

Q1-55. If F means output and F' means no output, then F can be represented by the digit _____ and F' by _____.

Q1-56. If F equals 1 × A, then F equals _____.

Q1-57. If F' equals 1 × A', then F' equals _____.

51

> **Your Answers Should Be:**
>
> **A1-55.** If F means output and F' means no output, then F can be represented by the digit 1 and F' by 0.
>
> **A1-56.** If F equals 1 × A, then F equals **A**.
>
> **A1-57.** If F' equals 1 × A', then F' equals **A'**.

Logical Addition

The OR circuit shown in Fig. 1-48 provides an output when A OR B OR C is present. An inspection of the timing diagram reveals that each signal is either 0 volts or 6 volts. These two voltage levels can be represented by the digits 1 and 0, respectively. For this case, we can represent the 6-volt level by a logical 1 and the 0-volt level by a logical 0. Notice that when A OR B OR C is a logical 1, the output is then a logical 1. Since the inputs are a 1 at four different times (twice for A), the output F is a 1 at those four times.

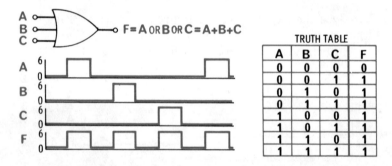

Fig. 1-48. Logical addition.

From the above information, we can construct the truth table for this three-input OR circuit. The truth table lists eight combinations starting with 000 and ending with 111. Since we have three inputs, (A,B,C) and each is a variable that can be either 0 or 1, there are two × two × two input combinations, which equal eight different possibilities. An analysis of the truth table indicates a 1 output (F) for seven of the eight input combinations. These seven conditions are summarized by F equals: 0, or 1. The OR circuit function is represented by the plus sign.

Negation

A NOT circuit negates, inverts, or complements an input signal. If the input to a NOT circuit is represented by the digit 1, the circuit provides a 0 output. Conversely, with a logical 0 input, the NOT circuit provides a 1 output.

Combining Logic Circuits

Up to this point, the basic logic circuits have been explained one at a time. You learned that the output from a logic circuit is determined entirely by the inputs to the circuit at that time. As shown in Fig. 1-49, a distinct difference exists between combinational circuits and sequential circuits. The outputs from *combinational circuits* depend entirely on their inputs and do not include memory elements. *Sequential circuits* include memory elements and logic circuits; their outputs are dependent on inputs and the state of the memory elements at that time.

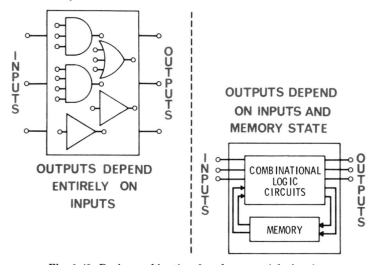

Fig. 1-49. Basic combinational and sequential circuits.

Q1-58. The output A + B is provided by an ____ circuit.

Q1-59. When signal W is applied to a NOT circuit, its output is then _____.

Q1-60. Circuits whose outputs depend entirely on their inputs are called _____ circuits.

53

> **Your Answers Should Be:**
> **A1-58.** The output A + B is provided by an **OR** circuit.
> **A1-59.** When signal W is applied to a NOT circuit, its output is then **W'**.
> **A1-60.** Circuits whose outputs depend entirely on their inputs are called **combinational** circuits.

COMBINATIONAL CIRCUITS

A circuit that combines two or more basic logic circuits (AND, OR, and NOT) is a combinational circuit, providing that it depends entirely on the input signals. You will learn about commonly used combinational circuits.

NOT OR Circuit

Although the NOT circuit has a single input, more than one signal can be applied to its input. Fig. 1-50 shows that when A OR B is applied to a NOT circuit, the output is NOT (A OR B). The NOT applies to everything within both parentheses. That is, the NOT of A is A'; the NOT of OR is AND; and the NOT of B is B'. Thus, NOT (A OR B) equals (A OR B)' which equals A' AND B', or simply A'B'. The NOT OR circuit is commonly referred to as a NOR circuit.

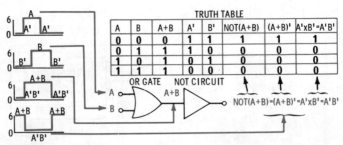

Fig. 1-50. NOT OR (NOR) circuit.

NOT AND Circuit

When A AND B are applied to the input of a NOT circuit, as shown in Fig. 1-51, the output is NOT (A AND B). The idea that everything within the parentheses is affected by NOT is extremely important. Then, NOT (A AND B) means the inverter circuit negates A to A', AND to OR, and inverts

B to B'. Therefore, NOT (A AND B) equals (A AND B)', which equals A' OR B'. The NOT AND circuit is usually referred to as a NAND circuit.

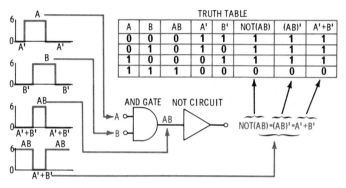

Fig. 1-51. NOT AND (NAND) circuit.

Exclusive OR Circuit

Another commonly used circuit is depicted in Fig. 1-52. This combined circuit requires two NOT circuits, two AND circuits, and a single OR circuit. An analysis of the truth table indicates A AND B' provides an output (1) for one out of the four input possibilities. Also, A' AND B provides an output under different input conditions. Then, the OR gate provides an output when A' AND B or A AND B' are present at the inputs to the exclusive OR circuit.

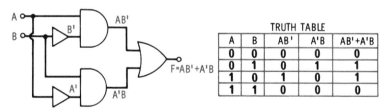

Fig. 1-52. Exclusive OR circuit.

Q1-61. A NOR circuit includes both _____ and _____ circuits.

Q1-62. A NAND circuit includes both _____ and _____ circuits.

Q1-63. An exclusive OR circuit consists of _____ OR, _____ AND, and _____ NOT circuits.

55

Your Answers Should Be:

A1-61. A NOR circuit includes both **NOT** and **OR** circuits.

A1-62. A NAND circuit includes both **NOT** and **AND** circuits.

A1-63. An exclusive OR circuit consists of **one** OR, **two**, AND, and **two** NOT circuits.

SPECIAL SEMICONDUCTOR CIRCUITS

Special semiconductor circuits contain solid-state devices that are not explained in other volumes of this transistor series. In Chapter 5 of this volume, you will learn about semiconductors as they are used in power supplies, communication, switching and control, and other circuits.

Extent of Coverage

Thousands of different semiconductors are available to equipment-design engineers. Each transistor differs from the others by certain characteristics such as operating temperatures, voltages, currents, gain, etc. An attempt to classify semiconductors by their applications necessarily limits the use of semiconductors in other areas.

Semiconductor circuits will be analyzed and the specific circuit application will be explained. The specific application will be representative and should not be interpreted as limited to that use alone.

Circuits consisting of discrete basic parts or integrated circuits and basic parts will be discussed. As shown in Fig. 1-53, printed-circuit cards (called modules if they are the plug-in type) can be used to interconnect the discrete basic parts, or the basic parts and the integrated circuits.

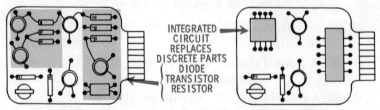

Fig. 1-53. Printed circuit and integrated circuits.

Additional Terminology and Abbreviations

Samples of the terms and symbols used in Chapter 5 follow. Fig. 1-54 shows the symbol and the typical method of connecting wires to a silicon controlled rectifier (scr) also called a *thyristor*. Notice that the symbol does not indicate that this device has two P-type and two N-type sections.

Fig. 1-54. Silicon controlled rectifier.

Another four-layer transistor device is shown in Fig. 1-55. The PNPN transistor is physically similar in appearance to the PNP transistor since both have three leads and the emitter symbols are identical. The base connection is called a gate and the collector is connected to an N-type section.

Fig. 1-55. PNPN transistor.

The symbol and basic construction of a field-effect transistor (fet) are shown in Fig. 1-56. Basic circuits using the field-effect transistor have been explained in Chapter 6 of Volume 2.

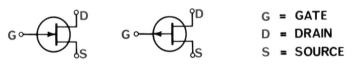

Fig. 1-56. Field-effect transistor.

Q1-64. The scr (thyristor) includes _____ P- and _____ N-type sections.

Q1-65. The base of a PNP transistor appears similar to the _____ of a PNPN transistor.

Q1-66. The three connections of a field-effect transistor are known as _____, _____, and _____.

> **Your Answers Should Be:**
> **A1-64.** The SCR (thyristor) includes **two** P- and **two** N- type sections.
> **A1-65.** The base of a PNP transistor appears similar to the **gate** of a PNPN transistor.
> **A1-66.** The three connections of a field-effect transistor are known as **drain, source,** and **gate**.

SUMMARY QUESTIONS

1. Digital circuits operate on numbers at high speeds in comparison to the pencil and paper method. Semiconductors are ideally suited for digital circuits because of their small size, efficiency, and reliability.
   a. Digital devices and circuits perform _____ with numbers.
2. The binary number system resembles the decimal system in that both are of the positional-notation type. The base of any positional-notation number system is determined by the number of digits used. In the decimal system, the base or radix is ten since ten digits are used; the base of the binary system is two.
   a. The digits __ and __ are used exclusively in the binary number system.
   b. In the Martian number system, the numbers 111 are _____ to earth inhabitants because the _____, positional notation, and weight (value) of the digits are unknown.
3. Logic circuits contain elements that behave like on-off switches and are sometimes called switching circuits. They are used in ADP equipment, digital computers, and real-time control systems. Business, industrial, scientific, and many other activities have increased their efficiency through the use of logic circuits. Some manufacturing processes would not be possible without real-time control systems.
   a. A system that provides an answer in time to do something about the problem is a (an) _____ _____ system.
   b. ADP means _____ _____ _____ .

4. Digital computer terms are common to logic-circuit expressions. The basic functional elements of a digital computer are the input and output elements, arithmetic, memory, and control elements.
   a. The computer programmer and the control element are linked by the _____ element.
   b. A "file cabinet" for both data and instruction words is the _____ or _____ element.
   c. Signals which direct the activities of the computer's other elements are generated by the _____ element.
5. The binary number system is ideal for the representation of two-valued conditions such as true or false, on or off, yes or no, etc. Because a transistor can be biased to cutoff (nonconducting) or to saturation (conducting), it is used extensively in logic circuits.
   a. A cut-off transistor is equivalent to an _____ switch. A conducting transistor is equivalent to a _____ switch.
   b. If a transistor that is biased to saturation is represented by the digit ___, then the same transistor that is biased to cut-off is represented by a _____.
6. Each digit in the decimal system increases by a value of ten from the least significant digit toward the more significant digit (units, tens, hundreds, etc.) In the binary system, each digit increases by a value of two times the number (one, two, four, eight, etc.).
   a. The decimal equivalent of binary 1001 equals _____.
   b. The decimal equivalent of binary 0011 equals _____.
   c. The decimal equivalent of binary 0001 equals _____.
7. To convert a decimal number to its binary equivalent: (a) divide the decimal number by two; (b) if the decimal number is even, record a 0 for the least significant digit of the binary number; (c) if the decimal number is odd, record a 1 for the least significant digit; (d) repeat the divide-by-two process until division by two is not possible; (e) for each process, record the 0 or 1.
   a. The binary equivalent of decimal 50 equals _____.
   b. The binary equivalent of decimal 63 equals _____.
   c. The binary equivalent of decimal 64 equals. _____.
8. To convert a binary number to its decimal equivalent (a) double the most significant binary digit; (b) if the

digit to the right of the most significant digit is 0, record a two; (c) if the digit to the right of the most significant digit is 1, record a three; (d) repeat the double-then-add process (double-dadd) for each binary digit; (e) for each process, record the results; (f) the number of binary digits minus one equals the required number of double-dadd processes.
  a. The decimal equivalent of binary 101 equals _____.
  b. The decimal equivalent of binary 1011 equals _____.
  c. The decimal equivalent of binary 101101 equals ____.
9. The rules for binary addition and subtraction are identical to those for decimal arithmetic. However, practice is necessary to become adept with the handling of zeros and ones only, rather than all the decimal digits. It is recommended that a beginner check binary arithmetic problems by decimal arithmetic.
  a. Decimal 3 equals binary _____. Binary 1011 plus decimal 3 equals _____.
  b. Decimal 5 equals binary _____. Binary 1111 minus decimal 5 equals _____.
10. The 1's complement in the binary number system is comparable to the 9's complements in the decimal system. Both 1's and 9's complements are also referred to as the radix-minus-one complements. Complementation has been explained so that you can understand logic circuits, rather than as a short-cut method of subtraction.
  a. The radix-minus-one complement of binary 1 equals _____.
  b. The 1's complement of 1010 equals _____.
11. Three binary digits can represent the digits zero through seven. However, to represent eight and nine, four binary digits are necessary. A code consisting of four binary digits to represent each decimal digit is called a Binary-Coded Decimal (BCD) code.
  a. A BCD code consisting of one group of four binary digits can represent _____ different numbers (include 0).
  b. With the excess-three code, _____ of all the possible combinations are not used.
12. Logic circuits require the representation of 1's and 0's by two valued signals. The two voltage levels are not

important in terms of 1's and 0's. Throughout this volume, the more positive voltage level represents a 1 and the less positive (or more negative) voltage level represents a 0. It should be understood that this convention is established to permit consistent presentation and is not binding on the design of logic circuits. Usually, throughout a given logic system, the representation of 1's by a relatively positive voltage level (positive logic), *or*, the representation of 1's by a relatively negative voltage level (negative logic) is consistent.

    a. If a signal at a voltage level of 6 volts represents a 1, then a 0 is represented by a voltage level _____ _____ than 6 volts.

    b. If a signal at a voltage level of −12 volts represents a 0, then a 1 is represented by a voltage level _____ _____ than −12 volts.

13. Flip-flops and monostable multivibrators are the two types of multivibrators that are used with logic circuits. They are two-state devices whose outputs are either 1 or 0. The two kinds of monostable multivibrators discussed are the one shot and the Schmitt trigger (ST). Flip-flops store bits while one-shot multivibrators and Schmitt triggers generate time delays.

    a. A type of multivibrator that _____ bits and is a two-state device is the _____-_____.

    b. Another type of multivibrator is the _____ multivibrator, which generates a time delay.

14. A flip-flop is the basic building block for binary and decade counters. Two successive input bits to a flip-flop provide a single output from the flip-flop. This divide-by-two characteristic of a flip-flop is called scaling and is used extensively in different kinds of scientific, commercial, and industrial counters.

    a. If 32 successive pulses are applied to five series connected flip-flops, then _____ pulse is provided from the output flip-flop.

    b. Divide-by-two is also called _____.

15. The basic processes performed by logic circuits are AND, OR, and NOT. An AND gate or circuit provides an output if and only if all inputs are simultaneously a logical 1. An OR gate provides an output if any one of the inputs is

61

a 1. The NOT circuit inverts, negates, or complements the input signal.
  a. An AND gate behaves like a _____ _____ switch-type circuit.
  b. If one of five inputs to an OR gate is a logical 1, its output is a logical _____.
  c. If the input to a NOT circuit is a logical 1, its output is a logical _____.
16. Symbolic logic, also called Boolean algebra, was developed to overcome difficulties resulting from the loose meaning of words. Symbols, rather than words, express relationships clearly and concisely. An understanding of symbols and the basic operation of logical multiplication, addition, and negation are essential. Logical multiplication is performed by an AND gate; logical addition by an OR gate; and negation by the NOT circuit.
  a. An AND gate performs logical _____.
  b. An OR gate performs logical _____.
  c. A NOT circuit performs _____.
17. Expressed in symbolic logic, multiplication performed by an AND gate is as follows:
  $A \times 1 = 1 \times A = 1A = A$
  $A \times 0 = 0 \times A = 0A = 0$
  $A \times B \times C = A \times C \times B = B \times C \times A = B \times A \times C = ABC = ACB \ldots$
  a. The expression A times A can be simplified to _____.
  b. The letter W times Y equals _____.
  c. The digit 0 times something equals _____.
18. Expressed in symbol logic, addition performed by an OR gate is as follows:
  $A + 1 = 1 + A = 1$
  $A + 0 = 0 + A = A$
  $A + B + C = A + C + B = C + A + B = C + B + A$
  a. The expression A plus 1 can be simplified to _____.
  b. The letter W or Y or Z equals _____ + _____ + _____.
  c. Something plus nothing equals _____.
19. Expressed in symbol logic, negation performed by a NOT circuit is as follows:
  NOT $1 = 1' = 0$
  NOT $0 = 0' = 1$
  NOT $F = F'$

a. If the input to a NOT circuit is W, then its output is _____.

b. The prime symbol (') means _____.

c. If the input to a NOT circuit is A', then its output is _____.

20. Combinational circuits consist entirely of AND, OR, and NOT circuits and their outputs depend entirely on their inputs. When the basic logic circuits (combinational circuits) are used with flip-flops or other storage (memory) devices, the circuits are called sequential. Three commonly encountered combinational circuits are the NOR, NAND, and exclusive OR circuits; all require the NOT circuit.

a. A NOR circuit includes both _____ and _____ circuits.

b. A NAND circuit includes both _____ and _____ circuits.

c. A combinational circuit does not contain _____ elements and their _____ depend entirely on their _____.

21. Due to the thousands of different semiconductor devices available, each differing by specific characteristics, any attempt to classify by application would limit the many possible uses of the semiconductor. Circuits consisting of both discrete parts and integrated circuits are used extensively in logic applications. Silicon controlled rectifiers (scr), PNPN transistors, and field-effect transistors (fet) are a few of the many special semiconductors used in power supplies, and other solid-state circuitry.

a. A sealed, self-contained unit containing two or more discrete parts is called an _____ circuit.

b. The scr is also called a _____.

c. The drain, source, and gate, are the three connections of a (an) _____ _____ transistor.

SUMMARY ANSWERS

1a. Digital devices and circuits perform **calculations** with numbers.
2a. The digits 0 and 1 are used exclusively in the binary number system.
2b. In the Martian number system, the numbers 111 are **meaningless** to earth inhabitants because the **base**, positional notation, and weight (value) of the digits are unknown.
3a. A system that provides an answer in time to do something about the problem is a **real-time control** system.
3b. ADP means **Automatic Data Processing**.
4a. The computer programmer and the control element are linked by the **input** element.
4b. A "file cabinet" for both data and instruction words is the **storage** or **memory** element.
4c. Signals which direct the activities of the computer's other elements are generated by the **control** element.
5a. A cut-off transistor is equivalent to an **open** circuit. A conducting transistor is equivalent to a **closed** circuit.
5b. If a transistor that is biased to saturation is represented by the digit **1**, then the same transistor that is biased to cut-off is represented by a **0**.
6a. The decimal equivalent of binary 1001 equals **9**.
6b. The decimal equivalent of binary 0011 equals **3**.
6c. The decimal equivalent of binary 0001 equals **1**.
7a. The binary equivalent of decimal 50 equals **110010**.
7b. The binary equivalent of decimal 63 equals **111111**.
7c. The binary equivalent of decimal 64 equals **1000000**.
8a. The decimal equivalent of binary 101 equals **5**.
8b. The decimal equivalent of binary 1011 equals **11**.
8c. The decimal equivalent of binary 101101 equals **45**.
9a. Decimal 3 equals binary **11**. Binary 1011 plus decimal 3 equals **1110**.
9b. Decimal 5 equals binary **101**. Binary 1111 minus decimal 5 equals **1010**.
10a. The radix-minus-one complement of binary 1 equals **0**.
10b. The 1's complement of binary 1010 equals **0101**.
11a. A BCD code consisting of one group of four binary dig-

its can represent **16** different numbers (zero through fifteen).

11b. With the excess-three code, **six** of all the possible combinations are not used. (They are: **0000, 0001, 0010, 1101, 1110,** and **1111**.

12a. If a signal at a voltage level of 6 volts represents a 1, then a 0 is represented by a voltage level **less positive** than 6 volts.

12b. If a signal at a voltage level of -12 volts represents a 0, then a 1 is represented by a voltage level **more positive** than -12 volts.

13a. A type of multivibrator that **stores** bits and is a two-state device is the **flip-flop**.

13b. Another type of multivibrator is the **monostable** multivibrator, which generates a time delay.

14a. If 32 successive pulses are applied to five series-connected flip-flops, then **one** pulse is provided from the output **flip-flop**.

14b. Divide by two is also called **scaling**.

15a. An AND gate behaves like a **series connected** switch-type circuit.

15b. If one of five inputs to an OR gate is a logical 1, its output is a logical **1**.

15c. If the input to a NOT circuit is a logical 1, its output is a logical **0**.

16a. An AND gate performs logical **multiplication**.

16b. An OR gate performs logical **addition**.

16c. A NOT circuit performs **negation**.

17a. The expression A times A can be simplified to **A**.

17b. The letters W times Y equals **WY**.

17c. The digit 0 times something equals **0**.

18a. The expression A plus 1 can be simplified to **1**.

18b. The letter W OR Y OR Z equals $\mathbf{W + Y + Z}$.

18c. Something plus nothing equals **something**.

19a. If the input to a NOT circuit is W, then its output is **W′**.

19b. The prime symbol (′) means **NOT**.

19c. If the input to a NOT circuit is A′, then its output is **A**.

20a. A NOR circuit includes both **NOT** and **OR** circuits.

20b. A NAND circuit includes both **NOT** and **AND** circuits.

20c. Combinational circuits do not contain **storage** elements and their **outputs** depend entirely on their **inputs**.

21a. A sealed, self-contained unit containing two or more discrete parts is called an **integrated** circuit.
21b. The scr is also called a **thyristor.**
21c. The drain, source, and gate are the three connections of a **field-effect** transistor.

2

Logic Circuits

What You Will Learn

This chapter describes the detailed functioning of circuit components (as shown in Fig. 2-1) for the basic logic elements covered in Chapter 1. You will learn how the basic logic elements, AND, OR, and NOT circuits, operate. The circuit function is explained by analysis of the effect of input signals on static circuit conditions; comparison of input- and output-signal levels, and then by the two-valued logical representations, 1's and 0's.

Two or more combined basic logic circuits, whose output is determined entirely by its inputs are called *combinational circuits*. The combinational circuits, NAND, NOR, Exclusive OR, and Inhibitor are also analyzed.

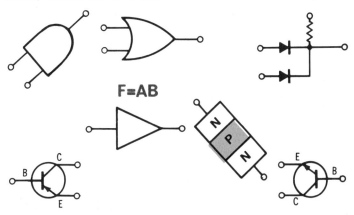

Fig. 2-1. Solid-state logic circuits.

SEMICONDUCTOR DIODES AND BIASING

The formation of positive-(P) type and negative-(N) type semiconductor materials which comprise the basic elements will be briefly discussed. A presentation of the directions of current versus electron flow will be followed by an explanation of conditions that permit a diode to be forward biased or reverse biased. We shall treat diodes as having either zero or infinite resistance since this volume is primarily concerned with two-valued, go/no-go circuits.

Semiconductor Materials

Germanium (Ge) and silicon (Si) are elements commonly used as basic semiconductor materials. When these elements are combined with small quantities of either acceptor or donor elements, P-type or N-type materials are formed. For example, if germanium is combined with an acceptor element such as gallium (Ga), a P-type material is formed. When Ge is combined with a donor element such as arsenic (As), an N-type material is obtained. Thus, negative (N-type) and positive (P-type) materials are formed by combining donor and acceptor elements respectively, with germanium or silicon. The N-type and P-type materials are used in the manufacture of diodes and transistors.

Current Versus Electron Flow

Two directions for current and electron flow are discussed in electronics. Conventional current is from a positive potential toward a negative potential. Electron flow is from a neg-

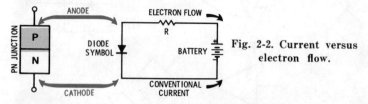

Fig. 2-2. Current versus electron flow.

ative potential toward a positive potential. This volume will not explain the advantages and disadvantages of adopting one explanation in preference to the other. The point is that you should be familiar with both explanations as shown in Fig. 2-2.

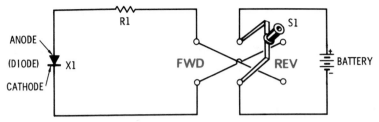

Fig. 2-3. Forward- and reverse-biasing.

Zero Bias, Forward Bias, and Reverse Bias

The biasing of a diode is shown in Fig. 2-3. Consider that the double-pole, double-throw (dpdt) switch has three positions: zero-bias, forward-bias, and reverse-bias.

When the dpdt switch is set to the zero-bias position, there is no current through the series-connected resistor and diode. With no current, there is no voltage drop (zero IR drop) across either the resistor or the diode. This constitutes a zero-bias condition.

When the dpdt switch is set to the forward-bias position, a negative potential is applied to the cathode of diode X1 and a positive potential is applied through resistor R1 to the anode of the diode. Under these conditions, the diode acts as a closed switch and looks to the remainder of the circuit as a zero resistance. Since the resistor and diode are series connected, the battery voltage is applied across R1.

When the dpdt switch is set to the reverse-bias position, a positive potential is applied to the cathode of the diode and a negative potential is applied through the resistor to the diode's anode. The diode acts as an open switch and looks to the remainder of the circuit as a very high (infinite) resistance. Since the resistor and diode are series connected, the battery voltage is applied across the high resistance of the diode.

Q2-1. Semiconductor materials are formed from germanium or silicon and _____ or _____ elements.

Q2-2. Conventional current is from _____ to _____ ; electron flow is from _____ to _____ .

Your Answers Should Be:

A2-1. Semiconductor materials are formed from germanium or silicon and **donor** or **acceptor** elements.

A2-2. Conventional current is from **positive** to **negative**; electron flow is from **negative** to **positive**.

DIODE LOGIC

Semiconductor diodes are widely used because of their low cost, small size, high speed, and long life.

Basic Diode Switch

A diode acts as a closed switch when its anode is positive with respect to its cathode, as shown in Fig. 2-4. When the input is at zero potential, the diode is forward biased and the output is also at zero potential. In effect, the input is connected to the output through a closed switch as shown.

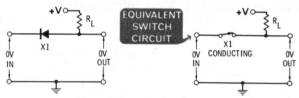

Fig. 2-4. Closed diode switch.

A diode acts as an open switch when it is reverse-biased, as shown in Fig. 2-5. When a positive input voltage (equal to or greater than +V) is applied to the diode's cathode, the diode is reverse-biased. In effect, we have an open switch between input and output terminals. Consequently, +V potential is applied to the output.

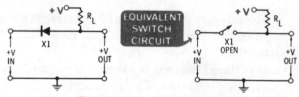

Fig. 2-5. Open diode switch.

Diode AND Gate

An AND gate provides an output only when all inputs are present simultaneously, as shown in Fig. 2-6. When zero voltage is applied to inputs A and B, both diodes are forward-biased. Both inputs are connected through the diodes to provide 0 volts at the output. When 0 volts is applied at either input A or B, 0 volts appears at the output through one of the forward-biased diodes.

When + voltages are applied to inputs A and B, the diodes are reverse-biased. Consequently, the + voltage is applied through the resistor to the output.

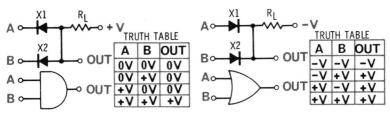

Fig. 2-6. Two-input diode AND gate.

Fig. 2-7. Two-input diode OR gate.

Diode OR Gate

An OR gate, as shown in Fig. 2-7, provides an output when one or more inputs are present. When negative voltages are applied to inputs A and B, the diodes are reverse-biased and a negative voltage is applied through the resistor.

When a positive voltage is applied to either or both inputs, the positive voltage appears at the output. The positive voltage forward-biases one or both diodes and is applied through the circuit to the output.

Q2-3. A diode is forward-biased when its anode voltage is _____ with respect to its cathode.

Q2-4. A diode AND gate provides zero volts at its output when one or more of its diodes are _____-biased.

Q2-5. A diode OR gate provides a positive voltage at its output when one or more diodes are _____-biased.

Your Answers Should Be:

A2-3. A diode is forward-biased when its anode voltage is **positive** with respect to its cathode.

A2-4. A diode AND gate provides zero volts at its output when one or more of its diodes are **forward-biased**.

A2-5. A diode OR gate provides a positive voltage at its output when one or more diodes are **forward-biased**.

TRANSISTOR TYPES AND CONFIGURATIONS

Different transistor types will be explained in terms of basic semiconductor materials that are also used in diodes. Symbols for transistor elements will be related to the corresponding materials. Also, the most important characteristics of different transistor configurations will be compared.

Transistor Types and Symbols

The next illustration, Fig. 2-8, shows that there are two general types of transistors, the NPN and PNP. Note that the symbols for the base and collector elements are identical for both transistor types, even though the symbols represent different basic materials. Also note that the arrow on the NPN emitter element is pointing away from the base to indicate the direction of conventional current. Usually, a schematic does not indicate whether a transistor is of the NPN or PNP type, except by the direction of the arrow on the emitter element. Therefore, an arrow on the emitter element that does *not* point toward the base indicates an NPN-type transistor.

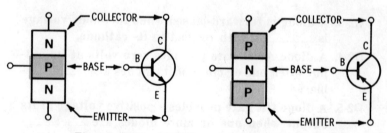

Fig. 2-8. NPN and PNP material and symbols.

Transistor Configurations

Transistors can be connected in three different circuit configurations, as shown in Figs. 2-9, 2-10, and 2-11. The most important characteristics of each circuit configuration are listed in Chart 2-1.

Chart 2-1. Circuit Configuration Characteristics

Characteristic	Common Base	Inverter	Emitter Follower
Power Gain	Yes	Yes	Yes
Voltage Gain	Yes	Yes	No
Current Gain	No (loss)	Yes	Yes
Input Impedance	Low	Medium	High
Output Impedance	High	Medium	Low
Input-Output Phase Relationship	In Phase	180° out of phase	In Phase

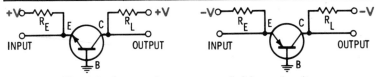

Fig. 2-9. Common-base or grounded-base circuit.

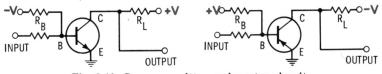

Fig. 2-10. Common-emitter or inverter circuit.

Fig. 2-11. Common-collector or emitter-follower circuit.

Q2-6. An arrow on the emitter element that does not point toward the base indicates a (an) _____-type transistor.

Q2-7. The characteristic that is common to all three circuit configurations is _____ _____.

Your Answers Should Be:

A2-6. An arrow on the emitter that does not point toward the base indicates an **NPN**-type transistor.

A2-7. The characteristic that is common to all three circuit configurations is **power gain.**

TRANSISTOR BIASING

Operating voltages applied to the input and output junctions determine whether the transistor will operate in the cutoff, active, or saturation region. Generally, logic circuits are biased to operate in the cutoff and saturation regions. Logic circuits are designed to make rapid transitions from one region through the active region to the other extreme.

Common-Base Circuit Biasing

The common-base circuits in Fig. 2-12 are reverse-biased through biasing resistor R_E with no signal present at the input. With the emitter-to-base junction reverse-biased, the transistor is cut off and there is practically no current. With no current through load resistor R_L, the collector-supply voltage is applied through R_L to the output.

With a signal of proper polarity applied to the emitter, the emitter-to-base junction is forward-biased and the transistor is turned on. High collector current is through the load resistor, the entire collector-supply voltage is dropped across R_L, and when the transistor is saturated (fully on), the collector voltage is zero.

It is important to understand that the collector-supply voltage for the NPN transistor must be positive; for the PNP transistor the supply voltage must be negative. This polarity can be easily identified as the center of the three letters that identify the transistor type.

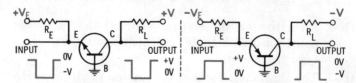

Fig. 2-12. Biasing the common-base circuit.

Biasing the Inverter Circuit

The inverter or common-emitter circuits shown in Fig. 2-13 are reverse-biased through base-biasing resistor R_B with no signal present at the input. This circuit provides a signal at its output when no signal is present at its input. When a signal of proper polarity is applied to the base, the transistor is turned on and the collector voltage goes to zero. Note that the input-signal polarity required to turn on the circuit is the same as the collector-supply voltage.

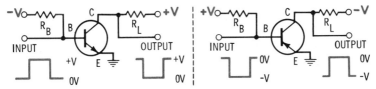

Fig. 2-13. Biasing the inverter circuit.

The common-collector or emitter-follower circuit is shown in Fig. 2-14. The voltage at the output, the emitter, follows the voltage that is applied to the input, the base. Unlike the inverter and common-base circuits, the emitter follower is operated in the region between cutoff and saturation called the *active region*. When the input to an emitter follower is at a 3-volt level, its output is also at 3 volts. The emitter follower is used for impedance matching, driving other circuits because of its power gain, and providing isolation between its high-impedance input and low-impedance output.

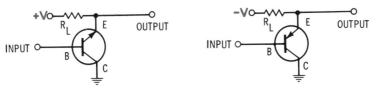

Fig. 2-14. Biasing the emitter-follower circuit.

Q2-8. An NPN transistor requires a (an) _____ collector operating voltage.

Q2-9. An NPN-type inverter circuit requires a positive going signal to _____-_____ its input junction.

> **Your Answers Should Be:**
> **A2-8.** An NPN transistor requires a **positive** collector operating voltage.
> **A2-9.** An NPN-type inverter circuit requires a positive going signal to **forward-bias** its input junction.

TRANSISTOR AND GATES

Transistors have reduced the need for additional amplifiers to overcome the energy loss associated with diodes. The AND gate has two or more inputs, a single output, and provides a signal output only when all input signals are present.

Emitter-Follower AND Gate

Fig. 2-15 shows three transistors connected as emitter followers performing the AND function. Load resistor R_L is common to all transistors. Inputs to this AND gate are either at a +5-volt or a −5-volt level. With three inputs to this circuit and each input two-valued, there are eight possible input combinations. All possible input combinations and their resultant outputs are listed below.

A	B	C	Output		A	B	C	Output
−5v	−5v	−5v	−5v		+5v	−5v	−5v	−5v
−5v	−5v	+5v	−5v		+5v	−5v	+5v	−5v
−5v	+5v	−5v	−5v		+5v	+5v	−5v	−5v
−5v	+5v	+5v	−5v		+5v	+5v	+5v	+5v

An analysis of Fig. 2-15 and the table above indicates that a −5-volt signal level at any one input forward-biases the base-to-emitter junction, that the transistor conducts

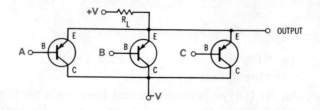

Fig. 2-15. Three-input transistor AND gate.

heavily, and the output is at a −5-volt level. When all inputs simultaneously receive +5-volt signals, the output is at a +5-volt level.

Direct-Coupled AND Gate

Fig. 2-16 shows a three-input AND gate using NPN transistors in an inverter configuration. When the input to any one of three transistors (Q1, Q2, and Q3) is at zero volts, that transistor is cut off and its collector-to-emitter resistance is high. In effect, an open circuit exists between the collector of transistor Q1 and ground. Positive supply voltage is applied through resistor R1 which turns on transistor Q4 by forward-biasing the base-to-emitter junction of Q4. The base bias on Q4 causes heavy collector current which effectively shorts Q4 collector to ground. Thus with one or more input transistors cut off, the collector of output transistor Q4 is at ground or zero volts.

An NPN transistor requires a positive voltage at its collector and a very small positive signal at its base (relative to the emitter) to turn the transistor on. When all inputs to transistors Q1, Q2, and Q3 are slightly positive, the current through resistor R1 is heavy. Effectively, the collector of transistor Q1 is at ground potential. Under these conditions, transistor Q4 is cut off and current through Q4 is very light. Thus, with all series-connected input transistors conducting heavily, output transistor Q4 current is effectively zero.

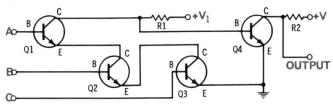

Fig. 2-16. Direct-coupled AND gate.

Q2-10. All inputs to an emitter-follower AND gate must be _____ to provide a positive voltage at its output.

Q2-11. When one input to the AND gate of Fig. 2-16 is at zero volts, transistor Q4 _____ conducting.

Q2-12. When all inputs to the AND gate of Fig. 2-16 are positive, transistor Q4 _____ conducting.

Your Answers Should Be :

A2-10. All inputs to an emitter-follower AND gate must be **positive** to provide a positive voltage at its output.

A2-11. When one input to the AND gate of Fig. 2-16 is at zero volts, transistor Q4 is conducting.

A2-12. When all inputs to the AND gate of Fig. 2-16 are positive, transistor Q4 **is not** conducting.

TRANSISTOR OR GATES

The circuits for two different OR gates are explained. The OR gate has two or more inputs, a single output, and provides an output when one or more of the input signals are present. It is suggested that explanations for the AND gates previously described be reviewed and compared with the OR gates. Note the similarity of circuitry, the different operating voltages required, and the polarity of input and output.

Emitter-Follower OR Gate

Fig. 2-17 shows three transistors connected as emitter followers to perform the OR function. An input signal to each of the bases (A, B, and C) can assume either a positive or negative 5-volt level. There are eight possible input combinations to this circuit, since it has three inputs and each is two-valued. These combinations and the resultant output for each combination are listed below.

A	B	C	Output	A	B	C	Output
−5v	−5v	−5v	−5v	+5v	−5v	−5v	+5v
−5v	−5v	+5v	+5v	+5v	−5v	+5v	+5v
−5v	+5v	−5v	+5v	+5v	+5v	−5v	+5v
−5v	+5v	+5v	+5v	+5v	+5v	+5v	+5v

An analysis of Fig. 2-17 and the table above indicates that a +5-volt signal at one or more inputs results in a +5-volt signal at the output of the circuit. A positive input forward-biases the associated base-to-emitter junction and that transistor conducts heavily. The heavily conducting transistor provides a current path from its input to the output of the circuit. The nonconducting transistors are reverse-biased by −5-volt signals at their inputs. The cut-off stages are isolated from the circuit's output and the stage that is conducting

Note that the circuit of Fig. 2-17 provides −5 volts at its output only when all inputs are also at −5 volts.

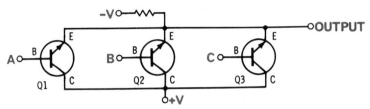

Fig. 2-17. Three-input transistor OR gate.

Direct-Coupled OR Gate

Fig. 2-18 shows a three-input OR gate using NPN transistors in an inverter configuration. A positive input to any one or more of the three transistors (Q1, Q2, and Q3) turns on the associated stage or stages. In effect, the base of transistor Q4 is at ground or zero potential and this stage is cut off. With transistor Q4 nonconducting, the voltage at the output rises to the positive supply voltage of V2, therefore, the circuit provides a positive output voltage when one or more inputs are positive. When all inputs are at zero volts, transistors Q1, Q2, and Q3 are cut off; Q4 conducts heavily and the output is at ground or zero potential.

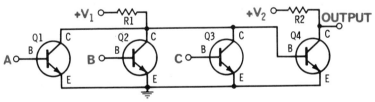

Fig. 2-18. Direct-coupled OR gate.

Q2-13. All inputs to an emitter-follower OR gate must be _____ to provide a negative output voltage.

Q2-14. A + V applied to one or more of the inputs to the OR circuit in Fig. 2-18 causes Q4 to be _____.

> **Your Answers Should Be:**
> **A2-13.** All inputs to an emitter-follower OR gate must be **negative** to provide a negative output voltage.
> **A2-14.** A +V applied to one or more of the inputs to the OR circuit in Fig. 2-18 causes Q4 to be **cut off**.

TRANSISTOR NAND GATE

The NAND gate is a basic building block frequently encountered in control systems, monitor and switching networks, and digital computers. This basic circuit is explained by input and output signal-voltage levels, then by the two-valued logical representations, 1's and 0's.

Direct-Coupled NAND Gate

Fig. 2-19 shows two NPN transistors series-connected to perform the NAND function. This illustration also shows that the output voltage level will never be the same as more than one of the input voltages at any given time. Notice that neither voltage has been assigned a truth value or logical representation of 1 or 0. Either transistor Q1 or Q2 will be conducting until +V is applied at both A and B at the same time. Stated differently, the voltage level at the collector of Q1 will be +V until the bases of both transistors assume voltage levels of +V.

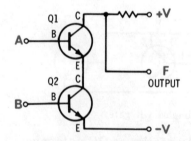

TRUTH TABLE		
A	B	OUTPUT
−V	−V	+V
−V	+V	+V
+V	−V	+V
+V	+V	−V

Fig. 2-19. Direct-coupled NAND gate.

In terms of the two-valued voltage levels, +V and −V, analysis of Fig. 2-19 reveals that when −V is present at either or both inputs the output F is at the +V level. Further, when both inputs are at +V, the output is at −V.

80

Positive Logic

When a positive voltage is represented by a logical 1 and a negative voltage by a logical 0, positive logic is used. A comparison of voltage levels to logical representations in terms of 1's and 0's is shown in Fig. 2-20. Absolute voltage levels are not assigned since they vary from one digital system to another. For example, one digital system may employ signal levels of +6 volts to represent a logical 1, and −6 volts to represent a logical 0. Another system may use +12 volts and 0 volts respectively.

In terms of logical 1's and 0's, the NAND circuit, shown in Fig. 2-19, provides a 1 output when one or both inputs are a logical 0. In addition, the NAND circuit's output is a logical 0 when both inputs are a logical 1.

A	B	OUTPUT
−V	−V	+V
−V	+V	+V
+V	−V	+V
+V	+V	−V

A	B	OUTPUT
−V	−V	+V
−V	+V	+V
+V	−V	+V
+V	+V	−V

A	B	OUTPUT
0	0	1
0	1	1
1	0	1
1	1	0

A	B	OUTPUT
1	1	0
1	0	0
0	1	0
0	0	1

(A) Positive logic.   (B) Negative logic.

Fig. 2-20. Logic representations.

Negative Logic

A positive voltage is represented by a logical 0 and a negative voltage by a 1 using negative logic. A comparison of relative voltage levels to 1's and 0's is shown in Fig. 2-20. As with positive logic, absolute voltage levels can be misleading. The important point to remember about negative logic is that the more negative (less positive) voltage is represented by a 1.

Q2-15. If +12 volts is represented by a logical 1, and + 6 volts a logical 0, _____ logic is being used.

Q2-16. When −6 volts is represented by a logical 1, and 0 volts a logical 0, _____ logic is being used.

81

> **Your Answers Should Be:**
> **A2-15.** If +12 volts is represented by a logical 1, and +6 volts a logical 0, **positive** logic is being used.
> **A2-16.** When −6 volts is represented by a logical 1, and 0 volts a logical 0, **negative** logic is being used.

TRANSISTOR NOR GATE

Another important building block frequently encountered in logic circuitry is the NOR gate. The circuit is explained by analyzing the effect of signal voltages on circuit characteristics, and then by the logical representations, 1's and 0's.

Direct-Coupled NOR Gate

Fig. 2-21 shows two NPN transistors connected as a direct-coupled NOR gate. Functionally, the NOR gate can be considered simply as an OR gate whose output is inverted.

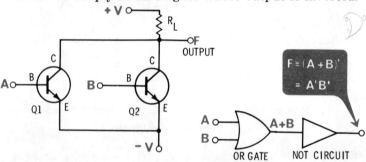

Fig. 2-21. Direct-coupled NOR gate and functional equivalent.

The signal voltage at inputs A and B and at output F can assume only one of two levels, +V or −V. A +V input signal at A or B forward-biases the base-to-emitter junction. With either or both transistors Q1 and Q2 conducting heavily, the supply voltage −V is connected through the low emitter-to-collector resistance to output F.

The output signal will assume the voltage level +V only when both transistors are cut off. Both transistors are non-conducting (cut off) when the base-to-emitter junctions are reverse-biased by signals at the voltage level of −V.

NOR Gate Voltage and Logic Levels

With a two-input gate, there are four possible combinations of input-voltage levels when each input can assume one of two levels. These combinations and the associated output-voltage levels are shown in Fig. 2-22 for OR and NOR gates using NPN-type transistors. This figure also shows the relationship of voltage levels to logic levels using positive logic. As an aid to recognizing positive logic, simply remember that a positive voltage is represented by a logical 1.

INPUT		OUTPUT	
A	B	OR	NOR
-V	-V	-V	+V
-V	+V	+V	-V
+V	-V	+V	-V
+V	+V	+V	-V

INPUT		OUTPUT	
A	B	OR	NOR
0	0	0	1
0	1	1	0
1	0	1	0
1	1	1	0

Fig. 2-22. OR and NOR gates, positive logic.

Fig. 2-23 shows the relationship between voltage levels and logic levels using negative logic. The voltage levels shown in Figs. 2-22 and 2-23 are identical. Notice that in comparison the logical levels are completely opposite in these illustrations. That is, wherever a logical 1 appears in Fig. 2-22, a logical 0 appears in Fig. 2-23. In effect, each signal in Fig. 2-22 has been passed through an inverter circuit to provide NOT signals, in terms of logic levels, to Fig. 2-23.

INPUT		OUTPUT	
A	B	OR	NOR
-V	-V	-V	+V
-V	+V	+V	-V
+V	-V	+V	-V
+V	+V	+V	-V

-V = LOGICAL 1
+V = LOGICAL 0

INPUT		OUTPUT	
A	B	OR	NOR
1	1	1	0
1	0	0	1
0	1	0	1
0	0	0	1

Fig. 2-23. OR and NOR gates, negative logic.

Q2-17. If both transistors of a positive logic NOR gate are cut off, the output will be a logical _____.

Q2-18. If both transistors of a negative logic NOR gate are conducting, the output will be a logical _____.

Q2-19. Using positive logic, a _____ voltage is represented by a logical 1.

> **Your Answers Should Be:**
> **A2-17.** If both transistors of a positive logic NOR gate are cut off, the output will be a logical 1.
> **A2-18.** If both transistors of a negative logic NOR gate are conducting, the output will be a logical 1.
> **A2-19.** Using positive logic, a **positive** voltage is represented by a logical 1.

TRANSISTOR EXCLUSIVE OR CIRCUIT

The exclusive OR circuit functionally consists of the three basic AND, OR, and NOT circuits. A special form of exclusive OR circuit is explained from both the voltage and logic levels standpoints.

Direct-Coupled Exclusive OR Circuits

Fig. 2-24 shows four PNP-type transistors connected as an exclusive OR circuit. The two-input OR gates previously discussed provide an output when one or the other or both inputs are present. Therefore, the previously discussed OR gates are called inclusive OR gates or circuits. In contrast, the exclusive OR circuit provides an output when one or the other input is present but not when both inputs are present at the same time.

A PNP-type transistor requires that a negative supply voltage be applied to its collector in order to reverse-bias the collector-to-emitter junction in an inverter (grounded-emitter) configuration. Also, the PNP transistor requires that a negative signal voltage be applied to its base in order to forward-bias the base-to-emitter junction and cause the transistor to conduct heavily. If a zero voltage is applied to the base of a PNP transistor used in a grounded-emitter configuration, that transistor is cut off (nonconducting).

The signals at inputs A and B can assume one of two voltage levels, 0 volts or $-V$ volts. When both inputs A and B are at the 0-volt level, transistors Q1 and Q2 are cut off.

The base and emitter of both transistors Q3 and Q4 are all at the same negative voltage level and cause Q3 and Q4 to be cut off. With all four transistors cut off the supply voltage $-V$ is applied through load resistor R_L to the output.

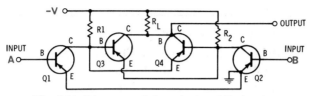

Fig. 2-24. Direct-coupled exclusive OR circuits.

When both inputs A and B are negative, Q1 and Q2 are turned on. The base and emitter of both Q3 and Q4 are at ground potential through the low resistance of Q1 and Q2. With no forward-bias on the base of Q3 and Q4, they are nonconducting and the negative supply voltage is applied through load resistor R_L to the output.

When Q1 is conducting and Q2 is cut off, the base of Q3 is at ground potential while its emitter is a negative voltage. Transistor Q3 is now reverse-biased and cut off while the emitter of Q4 is at ground and its base is at a negative voltage. These conditions cause Q4 to turn on. In effect, the output finds a low-resistance path through Q4 and Q1 to ground and the output is at ground potential.

When Q1 is cut off and Q2 is conducting, Q3 is conducting and Q4 is cut off. In this case, the output sees a low-resistance path to ground through Q3 and Q2.

Exclusive OR Voltage and Logic Levels

The following tables show the relationship of voltage levels to logic levels for the exclusive OR circuit.

A	B	Output		A	B	Output
−V	−V	−V		0	0	0
−V	0V	0V		0	1	1
0V	−V	0V		1	0	1
0V	0V	−V		1	1	0

Negative voltage equals logical 0. Zero voltage equals logical 1.

Q2-20. In the circuit shown in Fig. 2-24, Q2 and Q3 are _____ when input A is 0 and B is −V.

Q2-21. In the circuit shown in Fig. 2-24, Q3 and Q4 are _____ when inputs A and B are both 0V.

Q2-22. Functionally, the exclusive OR circuit is made up of _____, _____, and _____ gates.

> **Your Answers Should Be:**
> **A2-20.** In the circuit shown in Fig. 2-24, Q2 and Q3 are **conducting** when input A is 0 and B is −V.
> **A2-21.** In the circuit shown in Fig. 2-24, Q3 and Q4 are **cut off** when inputs A and B are both 0V.
> **A2-22.** Functionally, the exclusive OR circuit is made up of **AND, OR,** and **NOT** gates.

COMBINATIONAL CIRCUITS

Two or more combined basic logic circuits form a combinational circuit, if the output of the circuit is determined entirely by its inputs. The inhibitor circuit joins the NAND, NOR, and exclusive OR circuits in the family of basic logic building blocks.

Inhibitor Circuit

Fig. 2-25 shows the logic symbol for an inhibitor, another combinational circuit which consists functionally of an inverter and an AND circuit. A diagram for the inhibitor circuit will not be given since diagrams for the inverter and AND circuits have previously been presented and analyzed. In addition, it is probably more rewarding at this point to treat the relationship of basic logic circuits to each other, rather than analyze the functions of individual components (diode, transistors, etc.).

The purpose of an inhibitor circuit is to prevent the circuit from providing an output signal when an inhibiting signal is present. This circuit provides an output when all inputs are present except the inhibiting signal. Although only one input signal (A) and the inhibiting signal (B) are shown, this type of circuit can have several inputs.

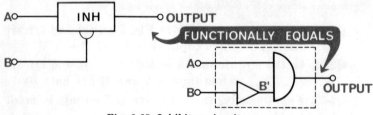

Fig. 2-25. Inhibitor circuit.

Truth Table for Inhibitor Circuit

As indicated by the truth table in Fig. 2-26, four combinations of input signals are possible. Assume that the presence of each input signal is represented by a logical 1 and its absence by a 0. Notice that signal A is connected directly to the input of the AND gate while the inhibiting signal B is inverted before being applied to the AND gate input.

The first row of the truth table indicates that signal A is a logical zero and signal B becomes B' or a 1 through the inverting action of the NOT circuit. The AND gate sees a 0 and 1 respectively for A and B; the circuit's output provides a 0 (no signal).

For row two of the truth table, the inputs of the AND gate see logical 0's and the circuit does not provide an output signal.

The third row indicates the AND gate sees logical 1's and the output is a logical 1. Notice that the inverter sees no signal at its input at this time.

For row four, the inputs of the AND gate see a 1 and a 0 respectively for A and B and the output is again a logical 0.

A	B	B'	OUTPUT
0	0	1	0
0	1	0	0
1	0	1	1
1	1	0	0

Fig. 2-26. Truth table for inhibitor circuit.

Q2-23. The circuit shown in Fig. 2-25 will provide an output when the inhibiting signal (is, is not) present.

Q2-24. If input A is + volts and input B is 0 volts the output of the inhibitor circuit is _____.

Q2-25. Functionally, the inhibitor circuit consists of a (an) _____ and a (an) _____ gate.

Your Answers Should Be:

A2-23. The circuit shown in Fig. 2-25 will provide an output when the inhibiting signal **is not** present.

A2-24. If input A is + volts and input B is 0 volts the output of the inhibitor circuit is **+volts**.

A2-25. Functionally, the inhibitor circuit consists of an **inverter** and an **AND** gate.

DUALITY OF LOGIC CIRCUITS

The dual nature of AND/OR circuits is explained in terms of positive and negative logic and their equivalency expressed in symbolic logic.

Positive Logic AND Gate

Fig. 2-27 shows the truth table for a two-input AND gate. For positive logic, the presence of a signal is represented by a logical 1 while a 0 represents the absence of a signal. This circuit provides an output only if signals A AND B are present at the inputs.

TRUTH TABLE

A	B	F	OUTPUT
0	0	0	NO
0	1	0	NO
1	0	0	NO
1	1	1	YES

1 DENOTES PRESENCE OF SIGNAL
0 DENOTES ABSENCE OF SIGNAL
$F = AB$

Fig. 2-27. Truth table for positive logic AND gate.

Negative Logic AND Gate

Fig. 2-28 shows the truth table for a two-input negative logic AND gate. A logical 1 represents the absence of a signal and a 0 represents the presence of a signal. The negative AND gate provides an output when either A OR B is present, or when both are present.

TRUTH TABLE

A	B	F	OUTPUT
0	0	0	YES
0	1	0	YES
1	0	0	YES
1	1	1	NO

0 DENOTES PRESENCE OF SIGNAL
1 DENOTES ABSENCE OF SIGNAL
$F = (A'B')' = A + B$

Fig. 2-28. Truth table for negative logic AND gate.

Positive Logic OR Gate

Fig. 2-29 shows the truth table for a two-input positive logic OR gate. A logical 1 represents the presence of a signal; a 0 represents the absence of a signal. The positive logic OR gate is identical to the negative logic AND gate since either circuit provides an output when A OR B or A AND B are present.

TRUTH TABLE

A	B	F	OUTPUT
0	0	0	NO
0	1	1	YES
1	0	1	YES
1	1	1	YES

1 DENOTES PRESENCE OF SIGNAL

0 DENOTES ABSENCE OF SIGNAL

$$F = A + B$$

Fig. 2-29. Truth table for positive logic OR gate.

Negative OR, Positive AND Circuits

The negative logic OR circuit performs the function identical to that of the positive logic AND circuit. Fig. 2-30 shows that either circuit provides an output only when all inputs are present simultaneously.

$$F = AB$$

TRUTH TABLE

A	B	OUTPUT	
		NEGATIVE LOGIC OR	POSITIVE LOGIC AND
0	0	YES	NO
0	1	NO	NO
1	0	NO	NO
1	1	NO	YES

Fig. 2-30. Truth table for negative OR, positive AND circuits.

Q2-26. A positive logic AND gate is comparable to a _____ logic _____ gate.

Q2-27. A negative logic AND gate is comparable to a _____ logic _____ gate.

Q2-28. The positive AND and negative OR function is expressed symbolically by F = _____.

Your Answers Should Be:

A2-26. A positive logic AND gate is comparable to a **negative** logic **OR** gate.

A2-27. A negative logic AND gate is comparable to a **positive** logic **OR** gate.

A2-28. The positive AND and negative OR function is expressed symbolically by F = **AB**.

LOGIC DIAGRAMS AND EQUATIONS

Logic circuits are frequently explained through the aid of logic block diagrams and symbolic logic equations. Logic diagrams are short-cut substitutes for schematics.

AND Gate Logic and Equations

Fig. 2-31 shows a two-input logic diagram and its associated truth table, modified to explain logic equations. The convention is used throughout that a logical 1 or an unprimed letter represents the presence of a signal and a logical 0 or a primed letter represents no signal. For row 4, then, F = AB means the circuit provides an output signal (F) when input signals A AND B are both present.

However, the expression F = AB does not present the complete picture regarding all possible input signal combinations and the resultant output. Rows 1, 2, and 3 respectively mean the circuit *does not* provide an output signal (F = no output) when: (1) Signals A AND B are *not* present (A' AND B' = A'B'); (2) Signal A *is not* present AND signal B is present (A' AND B = A'B); (3) Signal A is present AND signal B *is not* present (A AND B' = AB'). Then, F' = A'B' OR A'B or AB' = A'B + AB'.

	INPUTS		OUTPUT	OUTPUT vs INPUT EQUATIONS
1	0=A'	0=B'	0 = F'	F' = A'B'
2	0=A'	1=B	0 = F'	F' = A'B
3	1=A	0=B'	0 = F'	F' = AB'
4	1=A	1=B	1 = F	F = AB

F' = A'B' + A'B + AB'

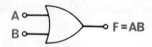

Fig. 2-31. AND gate logic and equations.

NAND Gate Logic and Equations

Fig. 2-32 shows a two-input NAND logic diagram and a modified truth table which is used to explain its logic equations. Row 4 of the truth table, $F' = A + B$, means that the NAND circuit *does not* provide an output when signals A AND B are both present at the inputs. The logic diagram suggests that signal AB, the output signal from the AND gate, is prevented from reaching the output of the NOT circuit. The NOT circuit provides a logical 0 output when signals A AND B are both present at its input. Under all other input combinations, the NOT circuit provides a logical 1 output since its input is a 0.

An analysis of row 1 indicates that the NAND gate provides an output when A' OR B' is NOT present at the input. An investigation of row 2 reveals that when signal A *is not* present (A') the circuit provides a 1 output. In comparing rows 1 and 2, it can be stated that neither B nor B' is essential at the input since A' is available.

A comparison of rows 1 and 3 indicates signal B *is not* present and is represented by B' in each case. Since B' is available at the input, neither A' nor A is necessary for the NAND circuit to provide an output signal.

Fig. 2-32. NAND gate logic and equations.

INPUTS		OUTPUT		OUTPUT vs INPUT EQUATIONS
A	B	AND	NAND	
0	0	0	1	F = A'+B'
0	1	0	1	F = A'+B
1	0	0	1	F = A +B'
1	1	1	0	F'= A +B

AND GATE   NOT CIRCUIT

A ○—⟩AB⟩○—○ F=(AB)'=A'+B'
B ○—

Q2-29. If input A to a two-input NAND gate is a logical 1, input B must be a logical _____ to satisfy the equation $F = A + B'$.

Q2-30. If the output of the AND gate, Fig. 2-32, is A'B, the output of the NAND gate is a logical _____.

91

> **Your Answers Should Be:**
>
> **A2-29.** If input A to a two-input NAND gate is a logical 1, input B must be a logical 0 to satisfy the equation $F = A + B'$.
>
> **A2-30.** If the output of the AND gate, Fig. 2-32, is $A'B$, the output of the NAND gate is a logical **1**.

TRUTH TABLES AND LOGIC

Truth tables present a convenient method of listing input signal combinations to logic circuits. With an understanding of operations performed by the basic logic circuits, input signal combinations can be evaluated and the resultant output can be accurately determined.

Stating the Problem

A problem is initially stated in words. For example, a simple black box is desired which contains three inputs and a single output. This logic package will provide an output signal when one of the three inputs is present, regardless of the states of the two remaining inputs. The output signal will also be present when a second signal is present while the third is not. The output signal will be dependent on inputs and shall not contain memory elements. (Memory elements will be covered in chapter 3.)

Constructing a Truth Table

For a simple logic problem, as stated above, $F = A + BC'$ applies. As an aid to explaining this expression, the truth table in Fig. 2-33 is constructed as follows:

1. Under columns A, B, and C, list all possible input signal combinations in ascending binary order. Thus, reading across row one, 000 equals binary 0; row 2 equals binary 1; etc. Notice row 8, 111 equals binary 7.
2. Under column C', list the 1's complement of C. That is, list a 1 under the C' column for each 0 under the C column; and a 0 under the C' column for each 1 under the C column.
3. Under the BC' column, list a 0 for each 0 under the B

and C' columns. For rows 3 and 7, list 1's under column BC' since B AND C' are both 1's.

A	B	C	C'	BC'	A+BC'= F
0	0	0	1	0	0
0	0	1	0	0	0
0	1	0	1	1	1
0	1	1	0	0	0
1	0	0	1	0	1
1	0	1	0	0	1
1	1	0	1	1	1
1	1	1	0	0	1

Fig. 2-33. Problem solving through a truth table.

4. Under column $A + BC' = F$, list 1's where A equals 1, and 1's where B AND C' equals 1's. List 0's for the remaining three rows.

Analysis of Truth Table

Each row of the truth table can be expressed as follows:

Row 1. $F' = A' + B'C'$ Row 5. $F = A + B'C'$
Row 2. $F' = A' + B'C$ Row 6. $F = A + B'C$
Row 3. $F = A' + BC'$ Row 7. $F = A + BC'$
Row 4. $F' = A' + BC$ Row 8. $F = A + BC$

A prime letter such as F' or A' means the signal is not present; an unprimed letter such as B or C means the signal is present. Rows 5 through 8 indicate signal A is present, so the second term is not essential for the output signal to be a 1. Row 3 of the truth table indicates the term BC' is a logical 1, so A is not essential for the output signal to be 1.

Q2-31. A neat, orderly method of arranging logic circuit input-signal data for evaluation is by use of _____ _____.

Q2-32. A two-input inhibitor circuit would have _____ possible input combinations.

> **Your Answers Should Be:**
>
> A2-31. A neat, orderly method of arranging logic circuit input-signal data for evaluation is by use of **truth tables**.
>
> A2-32. A two-input inhibitor circuit would have **four** possible input combinations.

SUMMARY QUESTIONS

1. Semiconductor materials are formed by combining acceptor or donor elements with either germanium or silicon. Adding acceptor elements forms a positive (P-type) material, adding donor elements forms a negative (N-type) material. The bias applied to a semiconductor diode will determine if the diode acts as an open or closed switch.
    a. When a small quantity of gallium is added to germanium, a (an) _____ semiconductor material is formed.
    b. A positive voltage applied to the anode, and a negative voltage to the cathode, will _____-bias the semiconductor diode.
2. Diode semiconductors are used extensively in logic circuits. The semiconductor acts as a closed switch when its anode is positive with respect to its cathode (forward-bias), and as an open switch when its cathode is positive with respect to its anode (reverse-bias). The AND gate provides an output only when all inputs are present at the same time, however the OR gate provides an output when one or more inputs are present.
    a. Both diodes of an AND gate (Fig. 2-6) must be _____-biased to provide an output.
    b. A forward-biased diode has _____ resistance, but a reverse-biased diode has _____ resistance.
    c. When one or more of the diodes of an OR gate (Fig. 2-7) are forward-biased, the output will be _____ voltage.
3. The two general types of transistors are the PNP and NPN. The schematic symbols are the same for both

with the exception of the arrow on the emitter element which points toward the base for the PNP and away from the base for the NPN. The three common transistor circuit configurations are the common base, common emitter (inverter), and the emitter follower.

   a. The arrow on the emitter of a PNP transistor points _____ the base.

   b. All three transistor circuits provide a power gain, but the _____ _____ and _____ circuits provide a voltage gain.

   c. A current gain is provided by the _____ and _____ _____ circuits.

4. Transistor bias, or the operating voltages, determine whether the transistor will be cut off or conducting. When the emitter-to-base junction is reverse-biased the transistor is cut off, when forward-biased the transistor will conduct. To have conduction the collector of the NPN must be positive, with respect to the base, but the PNP collector must be negative with respect to the base.

   a. Transistor logic circuits generally operate in the regions of _____ and _____.

   b. The input-signal voltage applied to a NPN inverter circuit must be _____ to cause the transistor to conduct.

   c. The input-signal voltage applied to a PNP common base circuit must be _____ to cause the transistor to conduct.

5. Transistor AND gates must have all inputs present in order to provide an output signal. The PNP emitter-follower AND gate (Fig. 2-15) input signals all must be positive to reverse-bias all transistors and provide a positive output voltage. The NPN direct-coupled AND gate (Fig. 2-16) input signals all must be positive to reverse-bias the output transistor and provide a positive-output voltage.

   a. A transistor AND gate must have _____ input signals present to provide an output signal.

   b. A negative signal input to one transistor of a PNP emitter-follower AND gate will cause that transistor to _____.

6. Transistor OR gates will provide an output signal when

any one or more of the input signals are present. The NPN emitter-follower OR gate (Fig. 2-17) input signal, or signals must be positive to forward-bias the transistor (or transistors) and provide a positive-output voltage. The NPN direct-coupled OR gate (Fig. 2-18) will provide a positive output when one or more of the input signals are positive, reverse-biasing the output transistor.
   a. An NPN emitter-follower OR gate will provide a negative output only when all transistors are _____-biased.
   b. The output transistor of an NPN direct-coupled OR gate will conduct only when all three input transistors are _____-biased.
7. Transistor NAND (NOT AND) gates will provide an output signal of the opposite polarity of the input signals only when all input signals are present. When a more positive voltage represents a logical 1, and a more negative voltage (less positive) represents a logical 0, positive logic is used. When using negative logic, the more negative voltage is represented by a logical 1, and a more positive voltage (less negative) is represented by a logical 0.
   a. The NPN direct-coupled NAND gate will provide a logical 1 output when all transistors are _____.
   b. If logical 1 is represented by positive 6 volts and logical 0 by positive 3 volts, _____ logic is being used.
   c. If input A to an NPN direct-coupled NAND gate is a logical 1 and input B is a logical 0, the output will be a logical _____.
8. Transistor NOR (NOT OR) gates operate as OR gates whose outputs are inverted. An NPN direct-coupled NOR gate (Fig. 2-21) will provide a negative output voltage if one or more of the inputs are positive. The output-voltage level will be positive only when all inputs are of the opposite polarity (negative).
   a. If all inputs to an NPN direct-coupled NOR gate are positive, all transistors will be _____-biased.
   b. Using negative logic, if all inputs to a NOR gate are logical 1's, the output will be a logical _____.
   c. Using positive logic, if input A to a NOR gate is posi-

tive, and input B is negative, the output will be a logical _____.
9. The exclusive OR circuit provides an output when either one or the other input is present, but not when both inputs are present simultaneously. The exclusive OR circuit functionally consists of the AND, OR, and NOT circuits.
   a. If both inputs to the direct-coupled exclusive OR circuit (Fig. 2-24) are 0 volts, transistors Q1 and Q2 are _____, and transistors Q3 and Q4 are _____.
   b. The output under the conditions stated above would be a logical _____.
10. Two or more combined basic logic circuits, where the output is determined entirely by its input signals are called combinational circuits. Examples of combinational circuits are the inhibitor, the exclusive OR, the NAND and NOR circuits. The inhibitor circuit consists of the basic AND and NOT (inverter) circuits. The inhibitor circuit provides an output signal when all inputs are present *except* the inhibiting signal, and no output when the inhibiting signal is present.
    a. If the input to a NOT circuit is a logical 0, its output will be a logical _____.
    b. If the A input to an inhibitor circuit is a logical 1 and the B input (inhibiting signal) is a logical 0, output will be a logical _____.
    c. The function of the NOT circuit is to _____ its input signal.
11. The basic logic AND and OR circuits perform dual functions. The AND gate using positive logic performs the same function as the OR gate using negative logic. Similiarly, the positive logic OR gate performs the same function as the negative logic AND gate.
    a. The symbolic expression $F = AB$ represents the positive _____ and negative _____ function.
    b. Using positive logic, the presence of a signal is represented by a logical _____, its absence a logical _____.
    c. A positive logic OR gate is the same as a _____ logic _____ gate.
12. Logic diagrams simplify the schematics used to repre-

sent equipment using many identical logic circuits. Logic equations say, in a few letters, that which would take many words to express. By this time, the logic diagram for the basic AND, OR, and NOT circuits should be readily recognized. The convention most commonly used in logic formulas is that an unprimed letter represents the presence of a signal; and a primed letter represents the absence of a signal.

    a. The logic equation for "An output signal (F) is present when both input signals A and B are present" is _____.

    b. The logic equation for "The output is not present when input signal A is present but input signal B is not present" is _____.

13. In addition to logic diagrams and equations, the truth table is another convenient tool used to simplify logic circuit input-output information. All possible input signal combinations are listed in ascending binary order. The input-signal combinations can then be evaluated to accurately determine the output signal.

    a. Construct a truth table for a two-input AND gate.
    b. Construct a truth table for a two-input OR gate.

SUMMARY ANSWERS

1a. When a small quantity of gallium is added to germanium, a **P-type** semiconductor material is formed.
1b. A positive voltage applied to the anode, and a negative voltage to the cathode, will **forward-bias** the semiconductor diode.
2a. Both diodes of an AND gate must be **reverse-biased** to provide an output.
2b. A forward-biased diode has **low** resistance, but a reverse-biased diode has **high** resistance.
2c. When one or more of the diodes of an OR gate are forward-biased, the output will be **positive** voltage.
3a. The arrow on the emitter of a PNP transistor points **toward** the base.
3b. All three transistor circuits provide power gain, but the **common base** and **inverter** circuits provide voltage gain.
3c. A current gain is provided by the **inverter** and **emitter-follower** circuits.
4a. Transistor logic circuits generally operate in the regions of **saturation** and **cutoff**.
4b. The input-signal voltage applied to an NPN inverter circuit must be **positive** to cause the transistor to conduct.
4c. The input-signal voltage applied to a PNP common-base circuit must be **positive** to cause the transistor to conduct.
5a. A transistor AND gate must have **all** input signals present to provide an output signal.
5. A negative signal input to one transistor of a PNP emitter-follower AND gate will cause that transistor to **conduct**.
6a. An NPN emitter-follower OR gate will provide a negative output only when all transistors are **reverse**-biased.
6b. The output transistor of an NPN direct-coupled OR gate will conduct only when all three input transistors are **reverse**-biased.
7a. The NPN direct-coupled NAND gate will provide a logical 1 output when all transistors are **cut off**.
7b. If logical 1 is represented by positive 6 volts and logical

0 by a positive 3 volts, **positive** logic is being used.
7c. If input A to an NPN direct-coupled NAND gate is a logical 1 and input B is a logical 0, the output will be a logical **1**.
8a. If all inputs to an NPN direct-coupled NOR gate are positive, all transistors will be **forward**-biased.
8b. Using negative logic, if all inputs to a NOR gate are logical 1's, the output will be a logical 0.
8c. Using positive logic, if input A to a NOR gate is positive, and input B is negative, the output will be a logical 0.
9a. If both inputs to the direct-coupled exclusive OR circuit (Fig. 2-24) are 0 volts, transistors Q1 and Q2 are **cut off,** and transistors Q3 and Q4 are **cut off.**
9b. The output under the conditions stated above would be a logical 0.
10a. If the input to a NOT circuit is a logical 0, its output will be a logical **1.**
10b. If the A input to an inhibitor circuit is a logical 1 and the B input (inhibiting signal) is a logical 0 output will be a logical **1.**
10c. The function of the NOT circuit is to **invert** its input signal.
11a. The symbolic expression $F = AB$ represents the positive **AND** and negative **OR** functions.
11b. Using positive logic, the presence of a signal is represented by a logical **1,** its absence, a logical **0.**
11c. A positive logic OR gate is the same as a **negative** logic **AND** gate.
12a. The logic equation for "An output signal (F) is present when both input signals A and B are present" is $\mathbf{F = AB}$.
12b. The logic equation for "The output is not present when input signal A is present but input signal B is not present" is $\mathbf{F' = AB'}$.
13a. Construct a truth table for a two-input AND gate.

A	B	F
0	0	0
0	1	0
1	0	0
1	1	1

13b. Construct a truth table for a two-input OR gate.

A	B	F
0	0	0
0	1	1
1	0	1
1	1	1

3

Multivibrator Circuits

What You Will Learn

There are two major types of multivibrator circuits: free-running and driven. The free-running multivibrator oscillates by itself, while the driven multivibrator requires an external trigger before it will generate an output waveform. This chapter describes the detailed functioning of the circuit components (Fig. 3-1) of the following multivibrator circuits: the free-running or astable, the one-shot or monostable, and the flip-flop or bistable. In addition, logical flip-flop circuits and equipment applications of multivibrator circuits will be discussed.

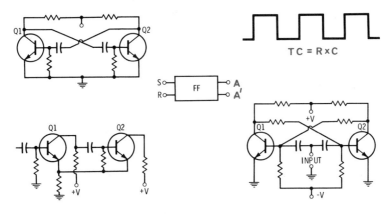

Fig. 3-1. Transistor multivibrator circuits.

103

WHY THE MULTIVIBRATOR IS REQUIRED

In many equipment applications, a rectangular voltage waveform is required. The collector-coupled free-running multivibrator circuit generates this rectangular voltage waveform. Before analyzing the multivibrator circuit, a brief description of the rectangular waveform and the need for the multivibrator will be discussed.

Description of a Rectangular Waveform

The rectangular waveform has four distinct characteristics which must be present. They are:

(1) A low steady level.
(2) A high steady level.
(3) An instantaneous rise from low to high steady level.
(4) An instantaneous drop from high to low steady level.

In Fig. 3-2 a rectangular waveform with steady levels of 0 volts and 12 volts is shown. It can be seen that the four requirements given above are present. The rectangular voltage

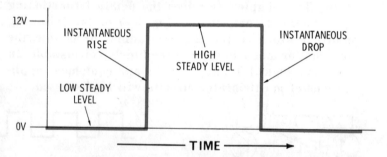

Fig. 3-2. Rectangular voltage waveform.

waveform has one steady value of 0 volts, an instantaneous rise from 0 volts to 12 volts, a steady level of 12 volts, and an instantaneous drop from 12 volts to 0 volts.

Development of a Rectangular Waveform

As shown in Fig. 3-3, a rectangular waveform can be generated by placing a switch in a common-emitter NPN amplifier circuit. When switch S1 is open, transistor Q1 does not conduct and the output is at the supply-voltage level.

When switch S1 is closed, transistor Q1 conducts and collector current is high, causing the output-voltage level to drop instantaneously to practically 0 volts, due to the voltage drop across collector load resistor (R2). As long as S1 is closed the output voltage will remain at the steady low level, but the instant that S1 is opened again, transistor Q1 is nonconducting, the collector current drops to 0, and the output voltage raises instantaneously to the high (supply-voltage) level. It remains at this level until S1 is closed again.

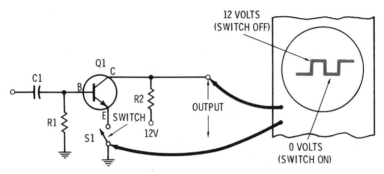

Fig. 3-3. Development of rectangular waveform.

Need for a Multivibrator

If a rectangular waveform can be generated by switching a transistor amplifier on and off, why the need for the multivibrator? The answer is time. The multivibrator circuit controls the period of time that the rectangular waveform is at both the high and low steady levels. This, in turn, determines the output frequency of the circuit. Also, it is difficult for a mechanical switch to perform the on and off switching functions at the very high speeds required, and maintain the symmetrical high and low time levels.

Q3-1. The output-voltage waveform of a free-running multivibrator is a (an) _____ waveform.

Q3-2. In addition to an instantaneous rise and drop, a rectangular waveform must have a (an) _____ level and a (an) _____ _____ level.

Q3-3. The multivibrator circuit controls the _____ of the rectangular output waveform.

Your Answers Should Be:

A3-1. The output voltage waveform of a free-running multivibrator is a **rectangular** waveform.

A3-2. In addition to an instantaneous rise and drop, a rectangular waveform must have a **high steady** level and a **low steady** level.

A3-3. The multivibrator circuit controls the **time** of the rectangular output waveform.

COLLECTOR-COUPLED MULTIVIBRATOR

A schematic used for analysis may look different from an equipment schematic. The following discussion will explain this difference.

Common Circuit Appearance

The schematic in Fig. 3-4 shows the common appearance of the collector-coupled free-running multivibrator found on equipment diagrams. Fig. 3-5 shows the same circuit as will be used for discussing circuit operation. Electronically, there is no difference between the circuits shown in Fig. 3-4 and 3-5. Fig. 3-4 shows the coupling circuit from the collector of Q2 to the base of Q1 on the righthand side of Q1, while Fig. 3-5 shows this same network on the lefthand side. The collector load resistors also are located differently in Fig. 3-4. The same circuit theory applies to both schematics as the components and their connections are identical.

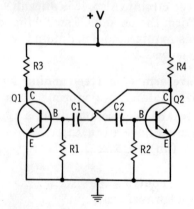

Fig. 3-4. Collector-coupled free-running multivibrator; common appearance.

Multivibrator Circuit Requirements

As you have previously learned, all oscillator circuits must have two characteristics: amplification and positive feedback. The circuit arrangement in Fig. 3-5 shows that the collector-coupled free-running multivibrator is nothing more than two common-emitter amplifiers, Q1 and Q2 coupled to each other to provide positive feedback. This circuit will generate the required rectangular waveform. In the circuit shown in Fig. 3-5, transistor Q2 will develop a rectangular output waveform when it is turned on and off. Transistor Q1 and the RC-coupling networks will determine the frequency of the rectangular waveform, or will determine how often Q2 will be turned on and off each second.

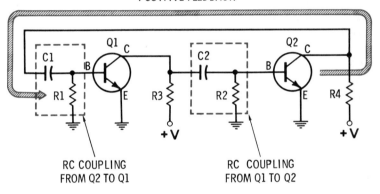

Fig. 3-5. A simple collector-coupled free-running multivibrator.

Q3-4. The two requirements of all oscillator circuits are _____ and _____ _____.

Q3-5. The collector-coupled free-running multivibrator circuit consists of two _____-_____ amplifiers with _____ feedback.

Q3-6. Transistor Q1 and the RC coupling networks will determine the output _____ of the multivibrator circuit.

Your Answers Should Be:

A3-4. The two requirements of all oscillator circuits are **amplification** and **positive feedback**.

A3-5. The collector-coupled free-running multivibrator circuit consists of two **common-emitter** amplifiers with **positive** feedback.

A3-6. Transistor Q1 and the RC coupling networks will determine the output **frequency** of the multivibrator circuit.

COLLECTOR-COUPLED MULTIVIBRATOR CIRCUIT OPERATION

One cycle of circuit operation will be explained in detail. Because the circuit is free-running, the explanation can start at any point. We will start with Q1 conducting and Q2 cut off. The voltage values used in the discussion are typical, and do not necessarily represent actual circuit values.

Q1 Conducting and Q2 Cut Off

The arrows in Fig. 3-6 show the direction of current through the circuit as Q1 conducts. The collector voltage of Q1 will be at its low value (nearly 0 volts) as the collector current of conducting Q1 is high. As transistor Q1 conducts, the discharge of capacitor C2 through resistor R2 will develop a voltage drop across R2 which is sufficient to reverse-bias transistor Q2. Transistor Q2 will remain cut off as long as the discharge of C2 through R2 is sufficient to reverse-bias transistor Q2.

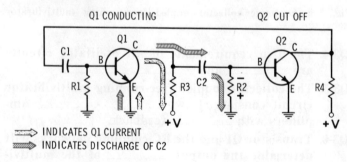

Fig. 3-6. Q1 conducting and Q2 cut off.

Switchover From Q1 Conducting to Q2 Conducting

After capacitor C2 has discharged through resistor R2 so that the voltage drop across R2 can no longer reverse-bias transistor Q2, transistor Q2 will conduct, as shown in Fig. 3-7. As transistor Q2 conducts, the collector voltage of transistor Q2 drops to its low value (nearly 0 volts), due to the high collector current of Q2. Capacitor C1, which was charged to the collector-supply voltage of Q2 (12 volts), now

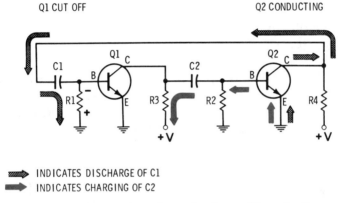

➡ INDICATES DISCHARGE OF C1
➡ INDICATES CHARGING OF C2

Fig. 3-7. Switchover from Q1 conducting to Q2 conducting.

has to discharge to the lower collector voltage (nearly 0 volts). The discharge path of C1 is through resistor R1, causing a voltage drop across resistor R1 which will reverse-bias transistor Q1 to cutoff. As transistor Q1 is reverse-biased below cutoff, its collector current will drop to zero, causing its collector voltage to increase to the supply-voltage value (12 volts). Capacitor C2 now charges to the high collector-voltage value of Q1 through the low emitter-to-base resistance of conducting transistor Q2.

Q3-7. As transistor Q1 conducts, transistor Q2 is _____ when capacitor _____ discharges through resistor _____.

Q3-8. The cutoff time of Q2 is determined by the time constant of resistor-capacitor _____ and _____.

Q3-9. The collector voltage of Q1 will be at its _____ value when Q2 is conducting.

Your Answers Should Be:

A3-7. As transistor Q1 conducts, transistor Q2 is **cut off** when capacitor C2 discharges through resistor **R2**.

A3-8. The cutoff time of Q2 is determined by the time constant of resistor-capacitor **R2** and **C2**.

A3-9. The collector voltage of Q1 will be at its **high** value when Q2 is conducting.

Q2 Conducting and Q1 Cut Off

Capacitor C2 charges quickly through the low emitter-to-base resistance of Q2, resulting in forward bias on Q2 (Fig. 3-8). With this bias, Q2 will conduct heavily, causing the collector voltage of Q2 to drop to its low value. The period of time that Q2 remains forward biased (conducting heavily) is controlled by the amount of time Q1 is cut off.

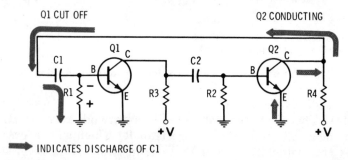

Fig. 3-8. Q2 conducting and Q1 cut off.

Return to Initial State

As soon as the discharge of capacitor C1 through resistor R1 decreases to a level insufficient to the reverse-bias Q1, it will permit Q1 to conduct; the collector voltage of Q1 will drop from its nonconducting level of 12 volts to its conducting level of nearly 0 volts. Capacitor C2 which was charged to the supply-voltage level of 12 volts will now have to discharge to the lower collector-voltage level, as shown in Fig. 3-9. The discharge of capacitor C2 through resistor R2 will develop a voltage drop across R2 which will reverse-bias Q2 to cutoff.

As transistor Q2 is cut off there is no collector current and the collector voltage of Q2 will increase to the supply-voltage level; it will remain at this level as long as the discharge current of capacitor C2 through resistor R2 is sufficient to reverse-bias Q2 to cutoff.

The length of time that transistor Q2 is cut off is determined by the time constant of resistor R2 and capacitor C2. The length of time that transistor Q1 is cutoff is determined by the time constant of resistor R1 and capacitor C1.

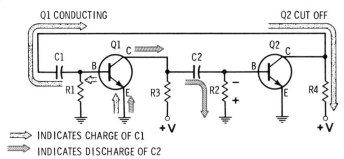

Fig. 3-9. Return to original condition; Q1 conducting, Q2 cut off.

Frequency Adjustment

The two RC networks, R1-C1 and R2-C2, will determine the frequency of the rectangular waveform of the multivibrator. The method most commonly used to vary the frequency of the multivibrator circuit is to change the time constant of the RC networks. Increasing the time constant decreases the frequency as the transistors are turned on and off a fewer number of times per second. The time constant can be increased by increasing the value of either R or C in the base circuits. Decreasing the time constant increases the frequency. The time constant can be decreased by decreasing the value of either R or C.

Q3-10. The period of time that Q2 remains conducting is determined by the time constant of _____ and _____.

Q3-11. If the time constant of an RC network is doubled, the output frequency will _____.

Q3-12. If the time constant of an RC network is halved, the output frequency will _____.

Your Answers Should Be:

A3-10. The period of time that Q2 remains conducting is determined by the time constant of **R1** and **C1**.

A3-11. If the time constant of an RC network is doubled, the output frequency will be **halved**.

A3-12. If the time constant of an RC network is halved, the output frequency will be **doubled**.

COLLECTOR-COUPLED MULTIVIBRATOR WAVEFORM ANALYSIS

To understand how the rectangular waveform is generated, each step of the circuit will be reviewed. Fig. 3-10 is keyed to each step of the circuit operation to develop the output rectangular waveform.

Step 1. Q1 conducting and Q2 cut off.
  (a) C1 is charged; the base of transistor Q1 is forward biased.
  (b) Q1 is conducting; Q1 collector voltage is at nearly 0 volts.
  (c) C2 is discharging through R2; Q2 is reverse-biased.

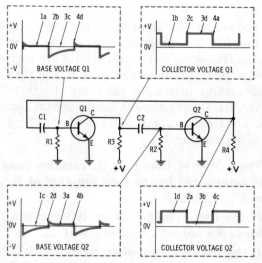

Fig. 3-10. Waveform analysis.

(d) Q2 is cut off; Q2 collector voltage is at 12 volts. C2 discharge current drops to a point which cannot keep Q2 cut off.

Step 2. Switchover from Q1 conducting to Q2 conducting.
    (a) Q2 starts to conduct; Q2 collector voltage drops to nearly 0 volts.
    (b) C1 starts to discharge through R1; Q1 is reverse-biased.
    (c) Q1 is cut off; Q1 collector voltage rises to 12 volts.
    (d) C2 charges; Q2 base voltage goes slightly positive.

Step 3. Q2 conducting and Q1 cut off.
    (a) C2 is charged; Q2 base is forward biased.
    (b) Q2 is conducting; Q2 collector voltage is at nearly 0 volts.
    (c) C1 is discharging through resistor R1; Q1 is reverse-biased.
    (d) Transistor Q1 is cut off; collector voltage of Q1 is at 12 volts. Capacitor C1 discharge current drops to a point which cannot keep Q1 cut off.

Step 4. Return to initial state.
    (a) Q1 starts to conduct; Q1 collector voltage drops to nearly 0 volts.
    (b) C2 starts to discharge; Q2 base voltage drops below cutoff.
    (c) Q2 is cut off; Q2 collector voltage rises to 12 volts.
    (d) C1 charges; Q1 base voltage goes slightly positive while C1 charges.

The action given in steps 1 through 4 is repeated many times each second, producing a rectangular waveform.

Q3-13. When capacitor C1 discharges through resistor R1, Q2 will be _____.

Q3-14. When Q1 is conducting, its collector voltage will will be at its _____ value.

Q3-15. When transistor Q2 is conducting the collector current of Q1 will be _____.

Your Answers Should Be:

A3-13. When capacitor C1 discharges through resistor R1, Q2 will be **conducting**.

A3-14. When Q1 is conducting, its collector voltage will be at its **low** value.

A3-15. When transistor Q2 is conducting the collector current of Q1 will be **zero**.

MONOSTABLE (ONE-SHOT) MULTIVIBRATOR

The one-shot multivibrator is a driven type, that is, one that requires an external trigger or input pulse for every output waveform. The normal output state of the one-shot with no input signal is its low-level or logical 0 state. When an input trigger is applied, the output is switched from its low level (logical 0) to its high level (logical 1). The circuit characteristics determine how long it will remain in the high level or 1 state.

Initial-State Operation

When normal operating voltage is applied to the one-shot multivibrator, there is no emitter-to-base bias on Q2, so transistor Q2 conducts as shown in Fig. 3-11. The collector current of Q2 goes through resistor R2, developing a voltage drop across R2 which reverse-biases Q1 to cutoff. As Q2 conducts and Q1 is cut off, capacitor C1 charges to the collector voltage of Q1 (supply-voltage value). The one-shot will remain in this condition; Q2 conducting and

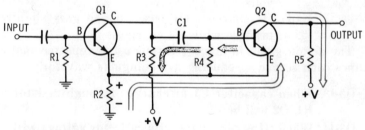

⟹ INDICATES C1 CHARGE PATH
⟹ INDICATES Q2 CURRENT PATH

Fig. 3-11. One-shot multivibrator; initial state.

Q1 cut off, due to the reverse bias across resistor R2, until a positive-trigger pulse is applied to the base of Q1.

Application of Trigger Pulse

When a positive-trigger pulse is applied, as shown in Fig. 3-12, the positive voltage applied to the base overcomes the emitter-to-base bias of Q1, causing Q1 to conduct. As Q1 conducts, its collector voltage drops to nearly 0 volts. Capacitor C1 must now discharge to the lower collector voltage of Q1. The discharge path of C1 through Q2 emitter-to-base

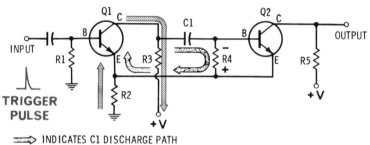

⟹ INDICATES C1 DISCHARGE PATH
▰▰▰▶ INDICATES Q1 CURRENT PATH

Fig. 3-12. One-shot multivibrator; application of trigger pulse.

resistor R4, as shown in Fig. 3-12. As capacitor C1 discharges through R4, the negative voltage developed across R4 reverse-biases Q2 to cutoff. The time constant of C1-R4 will determine how long transistor Q2 remains cut off with Q1 conducting. The positive-trigger pulse that was applied to the base of Q1 is of short time duration, as shown in Fig. 3-12. The only requirement of the trigger is that it starts the action of Q1 conducting. After the trigger pulse has dropped to 0 volts from its positive level, the circuit action of C1-R4 will determine the length of time that Q2 remains cut off.

Q3-16. The one-shot multivibrator must have a (an) _____ applied to switch the state of its output.

Q3-17. Emitter-to-base bias to cut off Q1 is developed across the common-_____ resistor.

Q3-18. The trigger applied to switch the state of the output waveform must be _____.

> **Your Answers Should Be:**
> **A3-16.** The one-shot multivibrator must have a **trigger** applied to switch the state of its output.
> **A3-17.** Emitter-to-base bias to cut off Q1 is developed across the common-**emitter** resistor.
> **A3-18.** The trigger applied to switch the state of the output waveform must be **positive.**

Return to Initial State

As soon as the discharge current of C1 through R4 has decreased to a value low enough to remove the reverse bias on Q2, transistor Q2 will start to conduct, as shown in Fig. 3-13. As Q2 starts to conduct, the collector current of Q2,

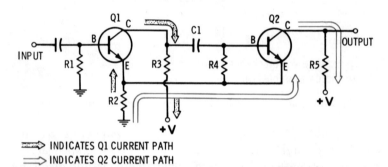

Fig. 3-13. One-shot multivibrator; return to initial state.

as well as that of Q1, will be through resistor R2. The increased current through R2 will reverse-bias Q1, causing Q1 to cut off. Q1 will remain cut off, due to Q2 collector current through resistor R2 until another positive trigger is applied to the base of Q1.

Frequency and Output-Pulse Duration

The trigger pulses determine the number of output pulses that the one-shot will provide. Stated differently, the pulse repetition rate of a one-shot is determined by the pulse repetition rate of the trigger pulses. However, the pulse duration or length of time that the output pulse of the one-shot will remain at the high (voltage) level is determined by the time constant of C1-R4. As shown in Fig. 3-14, each input trigger pulse causes Q1 to conduct and Q2 to cut off. If the time constant of C1-R4 is long, Q2 will remain cut off for a longer period of time than when the time constant is short. The time constant, therefore, determines how long Q2 collector voltage remains at the high level.

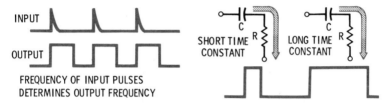

Fig. 3-14. Frequency and output pulse duration time.

Circuit Variations

One of the primary considerations of the one-shot is frequency stability. To achieve the stable output waveform required, the base resistor of Q2 is connected to the positive supply voltage, as shown in Fig. 3-15. The circuit operation is basically the same except that as capacitor C1 discharges through R4 it will discharge to a higher potential, using only the linear portion of the capacitor discharge curve. This results in an abrupt change from cutoff to conduction of Q2, giving improved frequency stability.

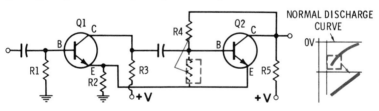

Fig. 3-15. One-shot multivibrator; circuit variation.

Q3-19. The output frequency of the one-shot is determined by the frequency of the _____ _____.

Q3-20. The time constant of _____ and _____ determines the time duration of the output pulse.

Q3-21. Connecting the grid resistor of the one-shot to the positive supply voltage will provide better _____ _____.

> **Your Answers Should Be:**
> **A3-19.** The output frequency of the one-shot is determined by the frequency of the **input trigger**.
> **A3-20.** The time constant of **C1** and **R4** determines the time duration of the output pulse.
> **A3-21.** Connecting the grid resistor of the one-shot to the positive supply voltage will provide better **frequency stability**.

ONE-SHOT WAVEFORM ANALYSIS

Each step of the circuit operation will be reviewed to show how the output waveform of the one-shot is developed. Fig. 3-16 is keyed to each step of the circuit operation to develop the output waveform.

Step 1. Q2 conducting; Q1 cut off.
 (a) Q2 conducts; Q2 collector voltage is at its low level.
 (b) Voltage across R2 is enough to reverse-bias Q1.
 (c) Q1 cut off; collector voltage is at its high level.
 (d) C1 charges to Q1 supply level; Q2 base is at zero bias.

Step 2. Application of trigger pulse.
 (a) Positive trigger pulse is applied to the base of Q1.
 (b) Q1 conducts; collector voltage drops to low level.
 (c) C1 discharges through R4, reverse-biasing Q2.
 (d) Q2 cut off; collector voltage at high level.
 (e) R2 voltage decreases due to cutoff of Q2.

Step 3. Q1 conducting, Q2 cut off.
 (a) Discharge of C1 through R4 keeps Q2 reverse-biased.
 (b) Collector voltage of Q2 remains at its high level.
 (c) R2 voltage remains at its low level due to Q1 emitter-to-collector current.
 (d) Q1 is conducting; its collector voltage is at its low level.

Step 4. Return to initial state.
 (a) Discharge of C1 through R4 drops, allowing Q2 to come out of cutoff.
 (b) Q2 conducts; Q2 collector voltage drops to its low level.

(c) R2 voltage increases due to Q2 collector current.
(d) Q1 is cut off due to emitter-to-base bias across R4; its collector voltage increases to its high level.

Step 5. Steady-state condition.
(a) Q1 is cut off; collector voltage remains at its high level.
(b) Capacitor C1 charges to the collector voltage of Q1; then Q2 remains at zero bias.
(c) Q2 continues to conduct; its collector voltage remains at its low level.
(d) R2 voltage remains at its high level to reverse-bias Q1.

Step 6. Application of next trigger pulse.
(a) As the next trigger pulse is applied to the base of Q1 the circuit action repeats.

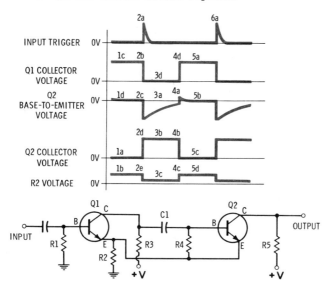

Fig. 3-16. One-shot waveform analysis.

Q3-22. When Q1 is cut off, the base voltage of Q2 is at _____ volts.

Q3-23. When Q2 is conducting, the voltage across R2 is at its _____ value.

Q3-24. At the instant that Q2 comes out of cutoff, collector voltage of Q1 will _____.

Your Answers Should Be:

A3-22. When Q1 is cut off, the base voltage of Q2 is at 0 volts.

A3-23. When Q2 is conducting, the voltage across R2 is at its **high** value.

A3-24. At the instant that Q2 comes out of cutoff, collector voltage of Q1 will **increase**.

ECCLES-JORDAN MULTIVIBRATOR

The Eccles-Jordan is a drive type of multivibrator; it requires an external trigger pulse at its input to obtain an output pulse. The Eccles-Jordan also is called a *trigger circuit* or, more commonly, a *flip-flop* because of the way the circuit operates. The flip-flop is the multivibrator circuit most commonly found in logic applications.

What the Eccles-Jordan Circuit Does

The Eccles-Jordan is a two-stage amplifier connected in such a way that when one stage conducts, the other stage cuts off. It is called a *bistable* circuit because it has two stable states of operation, as shown in Fig. 3-17. The circuit flips from one stable state to the other under the control of the input-trigger pulses. It takes two input pulses to produce a single square wave at the output. The first input pulse makes Q1 flip on and Q2 flop off, causing a sudden rise in the output-voltage waveform, which produces the leading edge of the output square wave. The output voltage remains at the higher level until the second input pulse is applied, causing Q2 to flip on and Q1 to flop off. As Q2 flips on, the

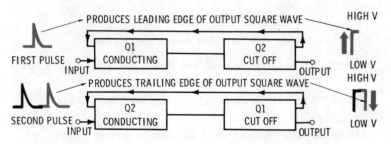

Fig. 3-17. How the Eccles-Jordan multivibrator works.

output voltage waveform suddenly drops to its low value, providing the trailing edge of the square wave.

Circuit Configuration

Fig. 3-18 shows the Eccles-Jordan circuit as it appears on most equipment schematics. Fig. 3-19 shows the same circuit simplified for the sake of analysis. Notice that the two circuits are identical except for the method of schematic

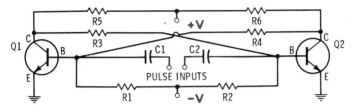

Fig. 3-18. Eccles-Jordan multivibrator; common appearance.

presentation. The circuit is a conventional two-stage amplifier with positive feedback, the only change being that resistors only are used to couple one stage to the other. A more symmetrical square wave is developed by eliminating the capacitors from the coupling circuits.

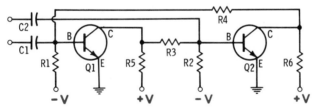

Fig. 3-19. Eccles-Jordan multivibrator; simplified appearance.

The circuit is arranged so that in its steady state (no input) one transistor is conducting and the other is cut off. When the proper input is applied, the cut-off stage conducts and the conducting stage is cut off. For each input pulse the two stages will reverse their previous states.

Q3-25. The Eccles-Jordan multivibrator is called a bistable circuit because it has _____ stable states.

Q3-26. A more symmetrical output waveform is obtained by use of _____ for the coupling.

121

> **Your Answers Should Be:**
> **A3-25.** The Eccles-Jordan multivibrator is called a bi-stable circuit because it has **two** stable states.
> **A3-26.** A more symmetrical output waveform is obtained by use of **resistors** for the coupling.

ECCLES-JORDAN CIRCUIT OPERATION

To understand circuit operation, several characteristics of the circuit shown in Fig. 3-20 must be seen. First, Q1 is in parallel with R3, R2, and −V2. Second, resistor R2 is the base resistor of Q2. Third, a drop in voltage across R1 lowers the positive voltage on the base of Q2, causing the −V2 to reverse-bias Q2 to cutoff. The same three statements apply to Q2 in parallel with R4, R1, and −V1. The voltage across R1 and −V1 control Q1.

Application of First Trigger Pulse

When plus voltages are applied to the circuit, assuming that Q1 conducts, the current through R2 and R3 drops because of the low resistance of conducting Q1 in parallel with them. This causes the positive base voltage on Q2 to drop, so that −V2 reverse-biases Q2 to cutoff. Since Q2 is cut off, it is not in parallel with R1 and R4, therefore current through R1 and R4 is high, keeping a positive voltage on the base of Q1 which overcomes −V1. This causes Q1 to conduct heavily.

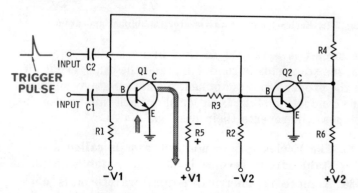

Fig. 3-20. Application of first trigger pulse.

A positive trigger pulse is applied to the base of Q2, the cut-off transistor. The positive pulse overcomes −V2, causing Q2 to conduct. As Q2 conducts, current through parallel resistors R1 and R4 drops suddenly. Therefore, the positive voltage on the base of Q1 drops, allowing −V1 to reverse-bias Q1 to cutoff. Q2 will remain conducting and Q1 cut off until a positive trigger pulse is applied to Q1 base.

Application of Second Trigger Pulse

When a positive trigger pulse is applied to the base of Q1, the cut-off transistor, it overcomes the reverse-bias voltage on Q1, causing Q1 to conduct. The same action explained for Q2 will now take place in Q1.

The positive trigger pulse affects only the cut-off transistor. A positive pulse applied to the conducting transistor will have no effect on that stage. In another circuit, the trigger is applied simultaneously to the base of Q1 and Q2 via a common input. The positive trigger pulse is applied to both base circuits, but affects only the cut-off transistor.

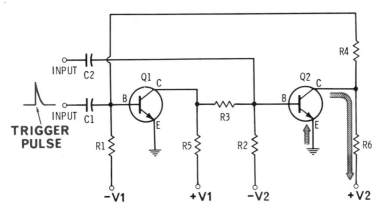

Fig. 3-21. Application of second trigger pulse.

Q3-27. The positive-input trigger pulse affects only the _____ transistor.

Q3-28. When a positive trigger pulse is applied to the base of Q2, it will cause the collector voltage of Q1 to _____.

Q3-29. When Q1 is cut off, the collector voltage of Q2 will be at its _____ value.

123

> **Your Answers Should Be:**
>
> A3-27. The positive-input trigger pulse affects only the **cut-off** transistor.
>
> A3-28. When a positive trigger pulse is applied to the base of Q2, it will cause the collector voltage of Q1 to **increase**.
>
> A3-29. When Q1 is cut off, the collector voltage of Q2 will be at its **low** value.

LOGICAL FLIP-FLOP

The flip-flop is an electronic switch capable of storing information. Specifically, it is used to store a binary digit or bit. This particular type of bistable device is called a *toggle*, a *binary*, and, very commonly, a *flip-flop*. Two or more binary-storage devices form a memory. We will not describe every kind of storage device nor every type of flip-flop. However, basic specifications for a binary-storage device will be provided so that you will also be able to apply the information to storage devices other than flip-flops.

Basic Specifications

The basic specifications for a binary-storage device are as follows:

1. Each device must have two distinct states and remain indefinitely in one state until it receives a signal to change to the other state.

2. There should be a means of determining the state that each storage device is in. Stated differently, means must be provided so that the truth value (1 or 0) of the bit stored in the device can be detected.

3. Means must also be provided to set or reset the device. When a bit whose truth value is 1 has been stored in the device, it is set. The device is reset or cleared when a 0 bit has been stored.

4. The storage device requires time (although very short) to change from one state to the other. Careful design of storage devices and associated circuitry ensures that the time required to make the transition from one state to the other is shorter than the time between input pulses to the

memory. Thus, each storage device has sufficient time between input pulses to reach one of the two states.

RS-Type Flip-Flop

A symbol for the RS-type flip-flop is shown in Fig. 3-22. RS stands for reset and set. The flip-flop is in the reset state after a signal has been applied to the R input. The A' output is a logical 1 and the A output is a logical 0 when the flip-flop is reset. Stated differently, the flip-flop has stored a 0 bit after it has been reset. The RS-type flip-flop will remain in the reset state until a signal is applied to the S input. That is, signals applied to the R input will not change the flip-flop's state when the flip-flop is reset.

The state of the flip-flop is changed from the reset state a short time after a signal is applied to input S. The A output is a logical 1 and the A' output is a logical 0 when the flip-flop is set. In other words, the flip-flop has stored a 1 bit after it has been set. Until a signal is applied to the R input, the RS-type flip-flop will remain in the set state.

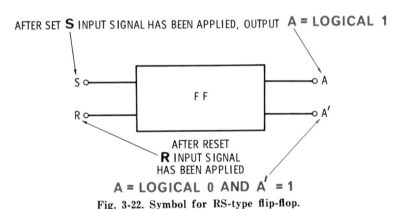

Fig. 3-22. Symbol for RS-type flip-flop.

Q3-30. A binary storage device is in the 1 state when the device is _____.

Q3-31. When a flip-flop has been set, it will be in the _____ state after another set signal has been applied.

Q3-32. When a storage device has stored a logical 0, it is said to be in the _____ state.

Your Answers Should Be:

A3-30. A binary storage device is in the 1 state when the device is **set**.

A3-31. When a flip-flop has been set, it will be in the **set** state after another set signal has been applied.

A3-32. When a storage device has stored a logical 0, it is said to be in the **reset** state.

SEQUENTIAL CIRCUITS

Combinations of logic circuits have been discussed in Chapter 2. When the basic logic circuit AND, OR, and NOT are combined, the output at any instant is a function of the inputs at that instant. A switching system must include a memory in addition to combination logic circuit in order to provide automation in communications, control, and computation. A switching system without memory is limited in application.

Sequential Circuit Types

Sequential circuits can be categorized into two types, *synchronous* and *asynchronous*. A synchronous sequential circuit is controlled by signals from a master clock or pulse generator. In contrast, the asynchronous sequential circuit starts an operation after receipt of a signal which indicates that a previous operation has been completed. Of course, the term asynchronous suggests that a clock is not utilized.

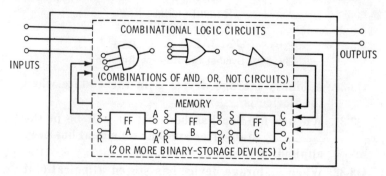

Fig. 3-23. Sequential circuit.

Sequential Circuit Applications

The memory portion of the sequential circuit shown in Fig. 3-23 enables a system to store information. Memory is defined as *a group of binary storage devices*. The memory, shown in Fig. 3-23, consists of RS-type flip-flops. It should be mentioned that other types of flip-flops are available and in common use. In addition, a memory that uses the magnetic properties of a material to store information could be used in place of the flip-flops.

The memory capability allows a person to place a telephone call by dialing a specific number. Switching and control circuitry record the call and automatically locate a path to reach the called party. Long distance direct dialing is made possible through memory and logic circuits.

Industrial processes are being monitored with greater frequency in order to take advantage of the increased speed and reliability provided by sequential circuits. To control an industrial process, information such as temperature, pressure, and density is supplied to a switching system. Through careful design, the switching system acts upon the industrial process to provide optimum results and tirelessly maintains the controlled quality.

Another important basic function performed by digital circuits is counting or computation. In essence, a counter is a device that records a number of events which have taken place. Usually, the input to an electronic counter consists of a serial string of pulses. The counter records and indicates the number of pulses which have occurred during a given period of time. Of course, memory is involved in this computation process since the counter must record the total number of pulses applied to its input.

Q3-33. Sequential circuits consist of combinational circuits plus _____.

Q3-34. Memory is a congregation of binary-storage devices used to _____ information.

Q3-35. A memory may use flip-flops or the _____ properties of a material.

> **Your Answers Should Be:**
> **A3-33.** Sequential circuits consist of combinational circuits plus **memory**.
> **A3-34.** Memory is a congregation of binary-storage devices used to **store** information.
> **A3-35.** A memory may use flip-flops or the **magnetic** properties of a material.

SUMMARY QUESTIONS

1. One of the primary applications of the multivibrator circuit is the generation of rectangular or square-wave voltages. The rectangular waveform has both low and high steady values, and instantaneous changes from one level to the other. The multivibrator circuit performs the function of generating the rectangular-voltage waveform much faster and with a more symmetrical shape than a mechanical switch can.
   a. The four requirements of a rectangular waveform are _____ and _____ steady levels and _____ rise and drop to these levels.
   b. The multivibrator generates a rectangular waveform by performing the function of a (an) _____ in an amplifier circuit.
2. The collector-coupled free-running multivibrator circuit consists of two transistor amplifiers with a feedback circuit from the output of each amplifier to the input of the other. The RC time constant of the coupling networks between the two amplifier circuits determines the output frequency of the rectangular waveform.
   a. The two requirements of the multivibrator circuit are _____ and _____ _____.
   b. The output waveform frequency is determined by the _____ _____ _____ of the coupling circuits.
   c. The collector-coupled multivibrator is one of the _____ type.
3. The collector-coupled free-running multivibrator circuit operation is such that as one transistor conducts, its collector voltage drops, causing a discharge of the coupling capacitor between stages through the base resistor, cut-

ting off that stage. The time constant of the RC-coupling network determines the length of time required for the resistor voltage to decrease to a value at which the cut-off transistor starts to conduct. As that transistor conducts, the drop in its collector voltage will cause a discharge of the other coupling capacitor through the base resistor, causing the previously conducting transistor to cut off. This conducting/cut-off action will repeat at a frequency determined by the two RC time constants.
   a. As transistor Q1 conducts, the collector voltage of Q2 is at its _____ value.
   b. As transistor Q2 conducts the base of Q1 is _____-biased.
   c. When Q1 is conducting the coupling capacitor between Q1 and Q2 is _____.
4. The method most commonly used to change the frequency of the output-voltage waveform is to vary the value of either R or C in the coupling networks. An increase in the value of either R or C will increase the time constant with a resultant decrease in frequency. A decrease in the value of either R or C will decrease the time constant with a resultant increase in frequency.
   a. If the value of R were doubled in the RC-coupling networks the output frequency would be _____ its original frequency.
   b. If the values of the capacitors in the RC-coupling networks were made one-half their original values, the output frequencies would be _____ their original frequencies.
   c. The output frequency of the multivibrator is 5000 kHz, the value of C is doubled, and R is halved; the output frequency will now be _____.
5. The one-shot multivibrator is driven. It requires an input trigger pulse to produce an output rectangular waveform. Each trigger pulse will provide one complete rectangular output pulse. The frequency of the one-shot is controlled by the repetition rate or frequency of the input trigger. The time constant of the coupling network varies the shape of the output waveform, but not the frequency. Circuit stability of the one-shot is improved by returning the base resistor of Q2 to the plus supply voltage.

a. The one-shot requires a (an) _____ input trigger pulse to produce an output waveform.
   b. If the input frequency of the trigger pulses is doubled, the output frequency will be _____.
   c. Frequency stability is improved when the base resistor is returned to the plus supply voltage because only the _____ portion of the coupling-capacitor's discharge curve is used.
6. The Eccles-Jordan multivibrator, or flip-flop as it is more commonly called, is a bistable device; it has two stable states. It requires an input pulse to change from one state to the other. The frequency or repetition rate of the input pulses determines how often the circuit changes states. The polarity of the input pulse must be such that it will change the state of the transistor to which it is applied. If the stage is cut off, a positive pulse must be applied to change the state to conduction. If the stage is conducting, a negative pulse must be applied to change the state to cutoff. In the flip-flop discussed in this chapter, positive polarity pulses are applied to the cut-off stage.
   a. A multivibrator that has two stable states is called a (an) _____ circuit.
   b. To change the state of a conducting transistor a (an) _____ pulse must be applied.
   c. The elimination of _____ in the coupling circuits of the flip-flop produces a more symmetrical output waveform.
7. The logical flip-flop provides the capability of storing information in the form of binary digits. A storage device must have the following specifications: (1) two distinct states; (2) a method of determining the state; (3) a means of changing the state; and (4) a rapid transition from one state to the other. The RS (Reset-Set) flip-flop is the one commonly found in logic applications. When the flip-flop has been set (by applying a pulse to the reset input) it is storing a binary 0. It will remain in this state until a pulse is applied to the set input, then it changes state and stores a binary 1 until another reset pulse is applied to the reset input.
   a. When a logical flip-flop is in the set state, it is storing a binary _____.

b. A combination of two or more storage devices forms a (an) ⎯⎯⎯⎯.
c. When a logical flip-flop is in its reset state, a (an) ⎯⎯⎯⎯ input is required to change its state.

8. A sequential circuit consists of a memory used in conjunction with combinational circuits. When a sequential circuit includes a pulse generator or clock, it is known as synchronous. An asynchronous sequential circuit cycles independently of clock pulses, in accordance with a previous operation that has been completed. A memory may consist of a group of flip-flops or a congregation of magnetic storage devices. Switching systems made up of sequential circuits enable the implementation of highly reliable and sophisticated communications media, industrial control and monitor systems, and automatic computation and recording devices.
    a. The purpose of a memory is to ⎯⎯⎯⎯ information.
    b. Combinational circuits plus memory are known as ⎯⎯⎯⎯ circuits.
    c. A synchronous sequential circuit makes use of a pulse generator or ⎯⎯⎯⎯ to synchronize each discrete action.

SUMMARY ANSWERS

1a. The four requirements of a rectangular waveform are **low** and **high** steady levels and **instantaneous** rise and drop to these levels.

1b. The multivibrator generates a rectangular waveform by performing the function of a **switch** in an amplifier circuit.

2a. The two requirements of the multivibrator circuit are **amplification** and **positive feedback**.

2b. The output waveform frequency is determined by the **RC time constant** of the coupling circuits.

2c. The collector-coupled multivibrator is one of the **free-running** type.

3a. As transistor Q1 conducts, the collector voltage of Q2 is at its **high** value.

3b. As transistor Q2 conducts, the base of Q1 is **reverse-biased**.

3c. When Q1 is conducting the coupling capacitor between Q1 and Q2 is **discharging.**

4a. If the value of R were doubled in the RC coupling networks, the output frequency would be **one-half** its original frequency.

4b. If the values of the capacitors in the RC-coupling networks were made one-half their original values, the output frequencies would be **double** their original frequencies.

4c. The output frequency of the multivibrator is 5000 kHz, the value of C is doubled, and R is halved; the output frequency will now be **5000 kHz**.

5a. The one-shot requires a positive input trigger pulse to produce an output waveform.

5b. If the input frequency of the trigger pulses is doubled, the output frequency will be **doubled.**

5c. Frequency stability is improved when the base resistor is returned to the plus supply voltage because only the **linear** portion of the coupling-capacitor's discharge curve is used.

6a. A multivibrator that has two stable states is called a **bistable** circuit.

6b. To change the state of a conducting transistor a **negative** pulse must be applied.
6c. The elimination of **capacitors** in the coupling circuits of the flip-flop produces a more symmetrical output waveform.
7a. When a logical flip-flop is in the set state, it is storing a binary **1**.
7b. A combination of two or more storage devices forms a **memory**.
7c. When a logical flip-flop is in its reset state, a **set** input is required to change its state.
8a. The purpose of a memory is to **store** information.
8b. Combinational circuits plus memory are known as **sequential** circuits.
8c. A synchronous sequential circuit makes use of a pulse generator or **clock** to synchronize each discrete action.

4

Waveshapers and Counters

What You Will Learn
This chapter describes circuits used to convert voltage waveforms to entirely different shapes. The detailed operation of limiters, squaring circuits, differentiators, and integrators will be explained by analysis of the effect of signals on static-circuit conditions, output signal waveshapes, and amplitudes.

In addition to waveshaping circuits, counter circuits will be explained in detail. Counter circuits, which are used for frequency division and control of actions in digital equipment at discrete time intervals, will be analyzed by using both logic levels and timing diagrams. Equipment applications of counter circuits will also be discussed.

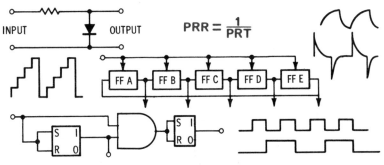

Fig. 4-1. Waveshapers and counters.

PULSE GENERATION

Pulses are used to represent information in communications, control, and computation equipment. They also provide the basic timing and control the sequence of operations within the equipment. An understanding of the circuits that generate and shape the pulses is essential.

Sine-Wave Oscillator

A sine-wave oscillator converts direct current (dc) to alternating current (ac). As shown in Fig. 4-2, a feedback oscillator consists of a frequency-determining network, an amplifier, and a feedback element. An oscillator can be identified by the type of frequency-determining network employed. Pulse counting and timing applications use oscillators with inductance-capacitance (LC) circuits and crystal-controlled circuits as the frequency-determining network.

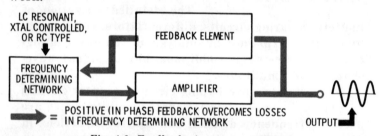

Fig. 4-2. Feedback sine-wave oscillator.

The frequency-determining network provides a single frequency signal to the amplifier input. The amplifier increases the power of the a-c signal and supplies part of the oscillator-output signal through the feedback element to the frequency-determining network. This feedback of energy is necessary to overcome resistive loss in the frequency-determining network. The feedback signal must be in phase with the signal from the frequency-determining element.

Pulse Waveforms

Distortion of a waveform is essential for the conversion from one waveshape to another. Waveshaping circuits are designed to deliberately introduce distortion so that one waveshape can be changed to another. For example, a waveshaping circuit can change a sine wave into a square wave. Fig. 4-3 shows several different nonsinusoidal waveforms that are common in digital equipment. These waveforms are alike in that they consist of harmonically related frequencies. In contrast, the sine wave represents a single frequency. As an example, a square wave is a sine wave of

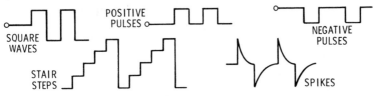

Fig. 4-3. Common pulse waveforms.

fundamental frequency plus sine waves that are the odd harmonics of that fundamental. The number of odd harmonics present, such as the 3rd, 5th, 7th, etc., determines the shape of the square wave. Usually 15 to 20 harmonics provide a square wave that is acceptable for digital applications.

Pulse Characteristics

A pulse is defined as a voltage or current that changes rapidly from one level of amplitude to another. The terms width, rise time, fall time (decay time), pulse repetition rate (prr), and pulse repetition time (prt) are commonly used to describe pulse characteristics. These characteristics are explained in Fig. 4-4.

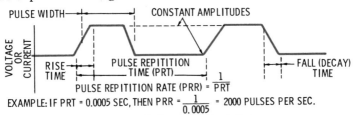

Fig. 4-4. Pulse characteristics.

Q4-1. A square wave with a prt of 1 microsecond has a prr of _____ pulses per second.

Q4-2. A pulse with short rise and fall times is rich in _____ _____ of the fundamental frequency.

> **Your Answers Should Be:**
>
> **A4-1.** A square wave with a prt of 1 microsecond has a prr of **1,000,000** pulses per second.
>
> **A4-2.** A pulse with short rise and fall times is rich in **odd harmonics** of the fundamental frequency.

AMPLITUDE-LIMITING CIRCUITS

Pulse timing and counter circuits require signals with specific waveshapes and discrete amplitude levels. Amplitude-limiting or clipping circuits change or distort waveshapes in order to provide signals required by pulse generation and regeneration circuitry.

Diode Limiter Circuit

A limiter circuit clips or limits a portion of the signal applied to it. Amplitude signal limiting can be performed by a resistor and diode, as shown in Fig. 4-5. This simple limiter circuit acts as a series-connected voltage divider on the input signal. Part of the input-signal voltage appears across the resistor and the remainder across the diode. Each positive half-cycle of the input signal sees the relatively high resistance of the resistor and the relatively low resistance of the forward-biased diode. Since the output signal from this limiter circuit is across the diode, the output is practically zero during the positive half-cycle.

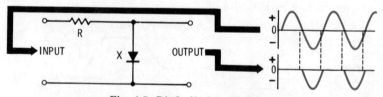

Fig. 4-5. Diode limiter circuits.

Each negative half-cycle of the input signal sees the relatively low resistance of the resistor and the relatively high resistance of the reverse-biased diode. Practically all of the negative half-cycle amplitude appears across the diode.

If the diode in Fig. 4-5 is reversed, negative half-cycles of the input signal will appear across the resistor. Positive half-cycles reverse-bias the diode and only positive signals will be present at the limiter-circuit output.

Squaring Circuit

Fig. 4-6 shows a circuit that limits the amplitude of both positive and negative excursions of a sine wave to provide a square-wave output. Notice that a positive 5-volt bias source is connected to the cathode of X1 and a negative 5-volt bias source is connected to the anode of X2.

The 5-volt bias source reverse-biases diode X1 until an input signal with an amplitude greater than 5 volts is applied to the anode. When the positive excursion of the input signal exceeds 5 volts, diode X1 is forward-biased and the diode resistance decreases to a relatively low value. Therefore, a positive half-cycle of input signal up to 5 volts appears across series-connected X1 and the positive 5-volt d-c source, while the portion of the positive input signal that is greater than 5 volts appears across the resistor. In this manner, the circuit limits the amplitude of the positive alternation at the circuit output to 5 volts.

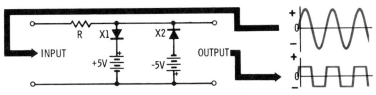

Fig. 4-6. Squaring circuit.

The negative 5-volt source connected to the anode of X2 reverse-biases this diode. The resistance of diode X1 remains high until the input signal applied to the cathode is more negative than −5 volts. When the input signal level overcomes the negative 5-volt level, diode X2 is forward-biased and its resistance drops to almost zero. Under these conditions, the input negative-going signal in excess of the bias voltage appears across the series-connected resistor.

Q4-3. That part of the input signal not appearing at the limiter circuit output, appears across _____.

Q4-4. A positive signal applied to the circuit shown in Fig. 4-5 _____-biases the diode X1.

> **Your Answers Should Be:**
>
> **A4-3.** That part of the input signal not appearing at the limiter circuit output, appears across **resistor R**.
>
> **A4-4.** A positive signal applied to the circuit shown in Fig. 4-5 **forward**-biases the diode X1.

DIFFERENTIATOR AND INTEGRATOR CIRCUITS

Differentiator and integrator circuits consist of resistance-capacitance (RC) networks that are used for changing the shapes of pulses. These circuits have one important common factor: the input signal is applied to a series combination of resistance and capacitance.

Differentiator Circuit

A differentiator circuit is used to convert a square-wave pulse to a peak wave, which is also referred to as a *trigger pulse, spike,* or *pip*. As shown in Fig. 4-7, an input pulse is applied across series-connected capacitor C and resistor R. The output waveform is taken from the junction of the two components to ground. Therefore, whatever voltage appears across the resistor constitutes the circuit output voltage. It

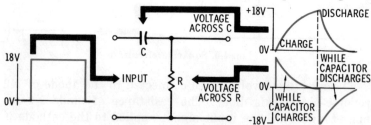

Fig. 4-7. Differentiator circuit.

is important to recognize that the sum of the voltages across the two components is equal to the applied voltage. For example, when an 18-volt input pulse is applied, the voltage across the resistor plus the voltage across the capacitor must equal 18 volts.

As the capacitor charges, the voltage across the resistor decreases. When the capacitor reaches a full charge, the voltage at the circuit output equals 0 volts. The width of the output waveform is dependent on the value of the circuit components R and C.

When the input pulse to the differentiator circuits drops to 0 volts, the fully charged capacitor immediately begins to discharge in a direction which is opposite to that of the capacitor charge direction. Note that the voltages across the resistor and capacitor are both tending toward a zero value.

Integrator Circuit

Fig. 4-8 shows that a square wave is converted to a triangular-shaped waveform by an integrator circuit. When an input pulse is applied, the output voltage (across the capacitor) slowly starts to increase, which indicates the capacitor is charging. Values of R and C are selected to provide a long time constant in comparison to the width of the input pulse. With properly selected values of R and C, the capacitor continues to charge but the voltage never equals the applied voltage level. As a result, the width of the rising output waveform is equal to the input pulse width.

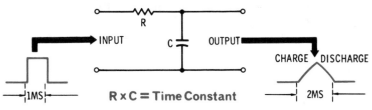

Fig. 4-8. Integrator circuit.

When the input pulse decays, the capacitor discharges and the output voltage decreases. Fig. 4-8 shows the width of the output waveform is twice the input pulse width.

Q4-5. In series-connected circuits, the sum of the voltages must equal the _____ voltage.

Q4-6. A capacitor in an integrator circuit does not _____ to the input-pulse voltage level.

Q4-7. A differentiator circuit provides an output voltage that instantaneously equals the _____ voltage.

> **Your Answers Should Be:**
> **A4-5.** In series-connected circuits, the sum of the voltages must equal the **applied** voltage.
> **A6-6.** A capacitor in an integrator circuit does not **charge** to the input-pulse voltage level.
> **A4-7.** A differentiator circuit provides an output voltage that instantaneously equals the **applied** voltage.

CASCADED BINARY CIRCUITS

A single flip-flop provides one output pulse for every two input pulses. For example, a flip-flop will provide 50,000 pulses per second from its output when 100,000 pulses per second are applied to the flip-flop input. With two cascaded or serially-connected flip-flops, 25,000 pulses per second are provided at the output when 100,000 pulses per second are applied to the input of the two-stage flip-flop circuit. Since each flip-flop divides the number of input pulses by two, the number of flip-flops becomes the exponent of two (2^n) in determining the divide-down ratio. When four flip-flops are cascaded, the circuit provides a divide-down ratio of 2^n or 2^4 or 16. That is, a circuit consisting of four cascaded flip-flops provides one output pulse for every 16 input pulses.

Block and Timing Diagram Analysis

The block and timing diagram for a serially-connected four-stage flip-flop is shown in Fig. 4-9. Assume that all flip-flops are reset prior to application of the first input pulse.

During the negative excursion of the first input pulse, flip-flop 1 changes to the set state. Flip-flop 1 remains set until the negative excursion of the second input pulse is present to reset flip-flop 1. Notice that at this point in time, two complete input pulses (two cycles) have caused flip-flop 1 to set and reset (one cycle). When flip-flop 1 is reset by the second input pulse, flip-flop 2 changes from the initial reset state to the set state. Two input pulses later in time, flip-flop 1 has been set, is now reset and flip-flop 2 is reset. When flip-flop 2 resets (after four input pulses are applied), flip-flop 3 is set for the first time. When flip-flop 4 is set (after sixteen

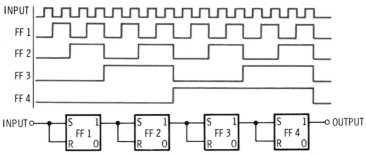

Fig. 4-9. Cascaded binary circuit, block and timing diagram.

input pulses have been applied), all flip-flops are again in the reset state.

Change-of-State Tabulation

The change of state that occurs in each of the four flip-flops relative to the presence of pulses at the input of the circuit is tabulated in Fig. 4-10. The tabulation provides a convenient means of determining the total number of input pulses that have been applied to the circuit. For example, when flip-flops 2, 3, and 4 are set and flip-flop 1 is reset, the number of input pulses that have been applied to the circuit equals binary 1110, or decimal 14.

INPUT PULSE	FF 4	FF 3	FF 2	FF 1	INPUT PULSE	FF 4	FF 3	FF 2	FF 1
0	0	0	0	0	9	1	0	0	1
1	0	0	0	1	10	1	0	1	0
2	0	0	1	0	11	1	0	1	1
3	0	0	1	1	12	1	1	0	0
4	0	1	0	0	13	1	1	0	1
5	0	1	0	1	14	1	1	1	0
6	0	1	1	0	15	1	1	1	1
7	0	1	1	1	16	0	0	0	0
8	1	0	0	0					

0 = RESET STATE   1 = SET STATE

Fig. 4-10. Tabulation of flip-flop states vs. input pulses.

Q4-8. Flip-flop 3 resets when the _____ input pulse terminates (refer to Fig. 4-9).

Q4-9. With flip-flop 4 reset, less than _____ input pulses of a 16-pulse group have been received.

Q4-10. With flip-flops 1 and 3 reset and flip-flops 2 and 4 set, _____ input pulses have been applied.

143

Your Answers Should Be:

A4-8. Flip-flop 3 resets when the **eighth** input pulse terminates (refer to Fig. 4-9).

A4-9. With flip-flop 4 reset, less than **eight** input pulses of a 16-pulse group have been received.

A4-10. With flip-flops 1 and 3 reset and flip-flops 2 and 4 set, **10** input pulses have been applied.

RING COUNTERS

A *ring counter* consists of a number of flip-flop stages that are connected to form a complete loop. Ring counters are used for frequency division and to control the actions within digital equipment at discrete times.

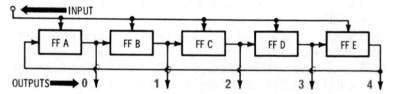

Fig. 4-11. Five-stage ring counter block diagram.

Fig. 4-11 shows a block diagram for a five-stage ring counter. All input pulses are applied to all flip-flop stages. However, each input pulse causes only two flip-flops to change state. One stage flips from set to reset while the next stage in the ring counter changes from reset to set.

Fig. 4-12 shows a timing diagram and state table for a five-stage ring counter. Analysis reveals that only one flip-

PULSE	FF A	FF B	FF C	FF D	FF E
NONE	1	0	0	0	0
1ST	0	1	0	0	0
2ND	0	0	1	0	0
3RD	0	0	0	1	0
4TH	0	0	0	0	1
5TH	0	0	0	0	0

LOGICAL **1** DENOTES FLIP-FLOP IS SET;
LOGICAL **0** DENOTES FLIP-FLOP IS RESET.

Fig. 4-12. Ring-counter timing diagram and state table.

flop is set at any one time and the remaining flip-flops are in the rest state. Notice that before application of the first input pulse, flip-flop A is set. The first pulse forces flip-flop A to reset and flip-flop B to set. The fifth input pulse applied to a five-stage ring counter forces flip-flop E to reset and flip-flop A to set.

Fig. 4-13 shows a three-stage ring counter that consists of three flip-flops and six AND gates. The states of the flip-flops before application of the first input pulse are: Flip-flop A (FF A) is set; FF B and FF C are reset.

The first input pulse resets FF A and sets FF B. The first pulse is gated with a logical 1 output from FF A, through AND gate 2, to the 0 input of FF A. Simultaneously, the input pulse is gated with the 1 output from FF A, through AND gate 3, to set FF B.

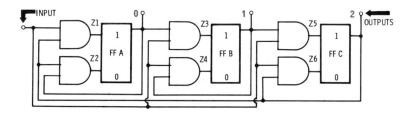

Fig. 4-13. Three-stage ring-counter logic diagram.

The second input pulse forces FF B to reset and FF C to set. FF B is reset through AND gate 4 and FF C sets through AND gate 5.

The third input pulse forces FF C to reset and FF A to set. FF C is reset through AND gate 6 and FF A is set through AND gate 1.

Q4-11. A ring counter with a capacity for counting from 0 through 9, requires _____ stages.

Q4-12. Before pulses are applied to the counter shown in Fig. 4-11, flip-flops B through E are _____.

Q4-13. When a third input pulse is applied to the counter shown in Fig. 4-13, flip-flops _____ and _____ change state.

> **Your Answers Should Be:**
> **A4-11.** A ring counter, with a capacity for counting from 0 through 9, requires 10 stages.
> **A4-12.** Before pulses are applied to the counter shown in Fig. 4-11, flip-flops B through E are **reset**.
> **A4-13.** When the third input pulse is applied to the counter shown in Fig. 4-13, flip-flops **C** and **A** change state.

DIVIDE-BY-FOUR PARALLEL COUNTER

Cascaded or serially-connected counters offer an advantage, in terms of the number of stages required, over a ring counter. However, a ring counter can switch states more rapidly than a serial counter because the stages of a ring counter are parallel to the input pulses. The two-stage parallel counter combines the advantages of a minimum number of stages and rapid counter responses to input pulses.

Divide-By-Four Logic

Fig. 4-14 shows two flip-flops and one AND gate connected as a divide-by-four counter. Flip-flops A and B are of the Reset-Set (RS) type and respond only to negative-going pulses. Therefore, only the negative spike of each serial input pulse is shown in the timing diagram in Fig. 4-15. Notice in Fig. 4-14 that the input pulse train is connected to both the set and reset inputs of flip-flop A and to one input of Z1.

Flip-flops A and B are initially reset before the application of input pulses. When the impact of the first negative-

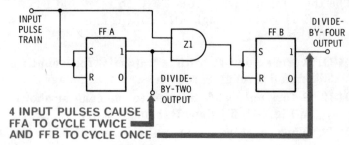

Fig. 4-14. Divide-by-four logic diagram.

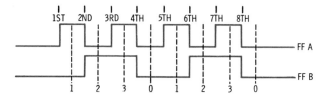

Fig. 4-15. Divide-by-four timing diagram.

going spike arrives at the input to the circuit, flip-flop A changes to the set state. The first input pulse is also present at one input of AND gate Z1. But, flip-flop B cannot change state at this time because Z1 sees only one negative-going input pulse since the 1 output from flip-flop A is going positive. Therefore, flip-flop A is set and flip-flop B remains reset after the first input pulse is applied.

When the second input pulse arrives, both flip-flops start to change state. Flip-flop A reverts to the reset state while flip-flop B flips to the set state. At this time, AND gate Z1 sees two negative-going input pulses.

While the 1 output from flip-flop A is going negative, the second input pulse is present and the coincidence of these signals is gated through Z1 to set flip-flop B. The second input pulse causes flip-flop A to rest and flip-flop B to set.

The third input pulse causes flip-flop A to set and has no impact on flip-flop B. That is, flip-flop B remains in the set state. AND gate Z1 sees the 1 output from flip-flop A going positive and inhibits the third input pulse from reaching the reset input of flip-flop B. After the third input pulse has been applied, both flip-flops are in the set state.

When the fourth input pulse is applied, flip-flops A and B start to reset. This negative-going input spike causes the 1 output from flip-flop A to go in the negative direction. Simultaneously, AND gate Z1 sees two negative-going inputs, which are gated to the reset input of flip-flop B. After the fourth input spike has terminated, the divide-by-four circuit has completed one full cycle.

Q4-14. Flip-flop A (Fig. 4-15) has changed state four times after _____ input pulses.

Q4-15. AND gate Z1 (Fig. 4-14) inhibits flip-flop B from changing state while flip-flop A is being _____.

147

Your Answers Should Be:
A4-14. Flip-flop A (Fig. 4-15) has changed state four times after **four** input pulses.
A4-15. AND gate Z1 (Fig. 4-14) inhibits flip-flop B from changing state while flip-flop A is being **set**.

DIVIDE-BY-THREE PARALLEL COUNTER

The two-stage parallel counter previously described can count from zero through three input pulses. The count in binary form for 8 input pulses is 01, 10, 11, 00, 01, 10, 11, 00. The same divide-by-four counter can be converted to a divide-by-three counter, which completes one cycle when three input pulses are applied.

Divide-By-Three Counter Logic Diagram

The logic diagram for a divide-by-three counter is shown in Fig. 4-16. This circuit consists of two flip-flops and two AND gates. Before input pulses are applied, both flip-flops are reset. AND gates Z1 and Z2 do not provide an output until both inputs to each gate are present at the same instant.

When the first negative-going input spike is applied, flip-flop A starts to flip from the reset to the set state. Before flip-flop A has had sufficient time to change state, AND gate Z2 inhibits the first pulse from reaching the set input of flip-flop B. At the same time, AND gate Z1 inhibits the first pulse from arriving at the reset input of flip-flop A. Regenerative feedback within flip-flop A completes the transition from reset to set a very short time after the input spike starts going positive.

When the second input spike arrives, flip-flop A is set and flip-flop B is reset. The 1 output from flip-flop A and the second input spike are gated through Z2 to the set input of flip-flop B. Before flip-flop B changes state, however, AND gate Z1 inhibits flip-flop A from resetting. A very short time after the second pulse is applied, flip-flop A and B are set.

When the third input spike is applied, flip-flops A and B are in the set state. The 1 output from flip-flop A, in time coincidence with the third input pulse, is gated through Z2 to the reset input of flip-flop B. However, before flip-flop B

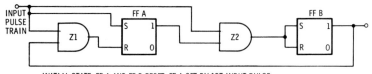

INITIAL STATE: FF A AND FF B RESET. FF A SET BY 1ST INPUT PULSE.
FF B SET BY 2ND INPUT PULSE AND "1" OUTPUT OF FF A.
3RD INPUT PULSE AND "1" OUTPUT OF FF B FORCES FF A TO RESET;
SIMULTANEOUSLY, 3RD INPUT PULSE AND "1" OUTPUT OF FF A FORCES FF B TO RESET.

Fig. 4-16. Divide-by-three counter.

changes state, the 1 output from flip-flop B is gated through Z1 by the third spike. Thus, flip-flops A and B are in the reset state a short time after the third pulse arrives.

Divide-By-Three Timing Diagram

Fig. 4-17 shows the timing diagram for the divide-by-three counter, depicted in logic diagram form in Fig. 4-16. In essence, the timing diagram provides an abbreviated explanation of the associated circuit. It is suggested that you reread the explanation of the divide-by-three counter while referring to the timing diagram instead of the logic diagram. In particular, attempt to relate the gating and inhibiting times to the presence of input pulses to the circuit. Also, observe the time required by each flip-flop to change from one state to the other.

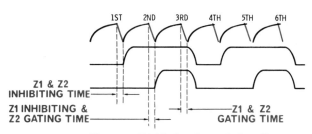

Fig. 4-17. Divide-by-three timing diagram.

Q4-16. Flip-flop A (Fig. 4-16) is inhibited from changing state by Z1 when flip-flop B is _____.

Q4-17. The divide-by-three counter changes state as follows: 01, _____, 00, 01, _____, 00, etc. (refer to Fig. 4-17).

Q4-18. AND gates Z1 and Z2 are both gating when the _____ input spike is present.

Your Answers Should Be:

A4-16. Flip-flop A (Fig. 4-16) is inhibited from changing state by Z1 when flip-flop B is **reset**.

A4-17. The divide-by-three counter changes state as follows: 01, **11**, 00, 01, **11**, 00, etc. (refer to Fig. 4-17).

A4-18. AND gates Z1 and Z2 are both gating when the **third** input spike is present.

DIVIDE-BY-FIVE PARALLEL COUNTER

A divide-by-five counter cycles once for each group of five input pulses. Three flip-flops must be arranged so that the present state of one flip-flop can be used to affect the next state of another flip-flop.

Changing State From Reset to Set

Fig. 4-18 shows the logic diagram for three flip-flops and five AND gates that are arranged to provide the divide-by-five function. Fig. 4-19 indicates the occurrence in time of input pulses or spikes relative to the set or reset state of each flip-flop.

The first negative-going spike starts flip-flop A into the set state. Before the transition of flip-flop A from the reset to the set state is completed, the input spike terminates. AND gates Z2 and Z4, respectively, inhibit flip-flops B and C from changing to the set state because the 1 output from flip-flop A is not present at this time. When flip-flop A is set, the state of this particular circuit can be expressed as 001.

The second input spike sets flip-flop B since this pulse is in coincidence with the 1 output from flip-flop A. The state of flip-flop A remains unchanged because AND gate Z1 in-

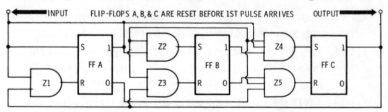

Fig. 4-18. Divide-by-five logic diagram.

hibits this flip-flop from being reset. AND gate Z4 inhibits flip-flop C from being set since the 1 output from flip-flop B is not present while the second input pulse is going negative. After the second input spike terminates, the state of this circuit can be expressed as 011.

The third input pulse, in coincidence with the 1 output from flip-flops A and B is gated through Z4 to set flip-flop C. The negative input spike terminates before flip-flop C is set so that flip-flops A and B remain in the set state. The state of the circuit after the third pulse can be expressed as 111.

Changing State From Set to Reset

When the fourth input pulse is applied, flip-flops A and B start to reset. This pulse is AND-gated with the 1 output from flip-flop C through Z1 and Z3, respectively, to reset flip-flops A and B. After the fourth input pulse, the circuit state can be expressed as 100.

The fifth input pulse, in coincidence with the 0 outputs from flip-flops A and B, is gated through Z5 to reset flip-flop C. After the fifth pulse, the circuit state can be expressed as 000, which means the three flip-flops are again in the reset state.

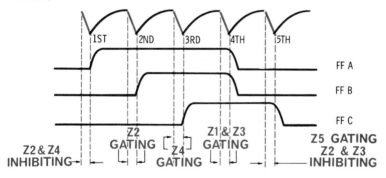

Fig. 4-19. Divide-by-five timing diagram.

Q4-19. The second input spike and the 1 output from flip-flop A (Fig. 4-18) sets flip-flop _____.

Q4-20. Fig. 4-19 indicates the circuit counts 001, 011, 111, _____, and 000 for five input pulses.

Q4-21. To reset FF C on the fifth input pulse, both FF A and FF B must be in the _____ state.

Your Answers Should Be:

A4-19. The second input spike and the 1 output from flip-flop A (Fig. 4-18) sets flip-flop **B**.

A4-20. Fig. 4-19 indicates the circuit counts 001, 011, 111, **100**, and 000 for five input pulses.

A4-21. To reset FF C on the fifth input pulse, both FF A and FF B must be in the **reset** state.

PARALLEL DECADE COUNTER

A divide-by-ten, or decade counter, cycles once for each group of 10 input pulses. Figs. 4-20, 4-21, and 4-22 show the logic diagram, state table, and timing diagram for a decade counter that counts in the binary number system.

As it was with previously discussed counters, all flip-flops are reset before the first input pulse is applied. Flip-flop A is set by each odd-numbered input pulse and reset by each even-numbered pulse.

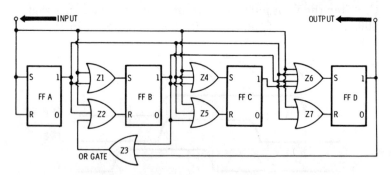

CIRCUIT COUNTS: 0001, 0010, 0100, 0101, 0110, 0111, 1000, 1001, 0000, 0001, 0010, ETC.

Fig. 4-20. Decade-counter logic diagram.

Flip-flop B is set by the second and sixth pulses, and the 1 output from flip-flop A. Inspection of Figs. 4-21 and 4-22 indicate that the fourth and eighth input pulses must reset flip-flop B. The 1 outputs from flip-flops A and B are AND-gated with the fourth and eighth input pulses to reset flip-flop B at these times. The 10th input pulse, in coincidence with the 1 outputs from flip-flops A and D, is applied

PRESENT STATE OF FLIP-FLOPS				INPUT PULSE ARRIVES	NEXT STATE OF FLIP-FLOPS			
D	C	B	A		D	C	B	A
0	0	0	0	1ST	0	0	0	1
0	0	0	1	2ND	0	0	1	0
0	0	1	0	3RD	0	0	1	1
0	0	1	1	4TH	0	1	0	0
0	1	0	0	5TH	0	1	0	1
0	1	0	1	6TH	0	1	1	0
0	1	1	0	7TH	0	1	1	1
0	1	1	1	8TH	1	0	0	0
1	0	0	0	9TH	1	0	0	1
1	0	0	1	10TH	0	0	0	0

0 DENOTES RESET STATE   1 DENOTES SET STATE

Fig. 4-21. Decade-counter state table.

through gate Z2 to prevent flip-flop B from changing to the set state.

Flip-flop C is set when AND gate Z4 provides a 1 output. The eighth pulse and the 1 outputs from flip-flops A and B cause flip-flop C to reset.

Flip-flop D is set when the eighth pulse is AND-gated with the 1 outputs from flip-flops A, B, and C. The 10th pulse and the 1 output from flip-flop A causes flip-flop D to reset.

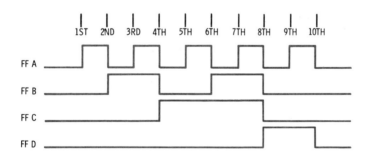

Fig. 4-22. Divide-by-ten timing diagram.

Q4-22. Flip-flop B is prevented from being set on the 10th pulse by feedback from the 1 output of flip-flop _____.

Q4-23. Flip-flop D is reset by the 10th pulse, which is AND-gated with the 1 output from flip-flop _____.

> **Your Answers Should Be:**
> **A4-22.** Flip-flop B is prevented from being set on the 10th pulse by feedback from the 1 output of flip-flop **D**.
> **A4-23.** Flip-flop D is reset by the 10th pulse, which is AND-gated with the 1 output from flip-flop **A**.

COUNTER APPLICATIONS

Counter applications can be grouped into two categories, direct and indirect counting. Direct counting is frequently used to control industrial processes. Counting of events through frequency division or scaling, and by comparing known frequencies with frequencies to be measured, fall into the category of indirect counting. The measurement of frequency, time, and speed usually requires more sophisticated circuitry than direct-counting applications.

Direct counting of events can be accomplished by an electronic counter with reliability where human operators fail because of fatigue or speed limitations. The events to be counted must be converted into electrical signals or pulses. Fig. 4-23 shows an example of direct counting. A photoelectric cell senses objects which are to be counted on a conveyer belt. Electrical signals from the photoelectric cell are shaped so that a counter can respond and provide a direct readout of the number of objects that pass a given point.

A preset counter, another form of direct counting, provides an output pulse or signal when a predetermined num-

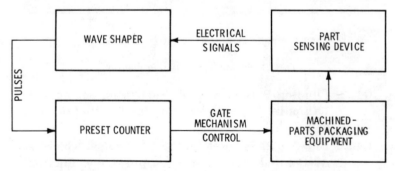

Fig. 4-23. Direct counting of events; block diagram.

ber of input pulses are applied. For example, a preset counter can be used with the packaging of small items such as machined parts. The counter, which is usually adjustable, is set to a certain number. When the predetermined number of machined parts are placed into a container, a gate mechanism, controlled by the output from the preset counter, directs the machined parts into the next container.

Fig. 4-24 illustrates the basic principle by which electronic counters are used to measure frequency. An input signal whose frequency is to be measured is applied through a gate to a counter. When the gate is opened for a known time interval, the frequency of the signal under measurement can be determined. For example, if the gate is kept open for 1 second, the counter provides the frequency under measurement in hertz. The time that the gate is kept open is determined by the known frequency of the crystal oscillator and the divide-down ratio of the dividers. To provide a gating time of 1 second, a 1-megahertz crystal oscillator must be shaped by the squarer, then applied through a 1-megahertz divider to the gate circuit.

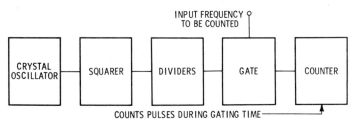

Fig. 4-24. Measurement of frequency.

Q4-24. The counting of events through the use of a preset counter results in one output pulse for a (an) _____ number of input pulses.

Q4-25. The gating time for a frequency to be measured by the indirect-counting method is determined by both the crystal oscillator and the _____ circuits.

Q4-26. If the gating time is 5 milliseconds and the electronic counter indicates 10 pulses are present during the gating time, the frequency under measurement is _____ hertz.

> **Your Answers Should Be:**
>
> **A4-24.** The counting of events through the use of a preset counter results in one output pulse for a **predetermined** number of input pulses.
>
> **A4-25.** The gating time for a frequency to be measured by the indirect-counting method is determined by both the crystal oscillator and the **divider** circuits.
>
> **A4-26.** If the gating time is 5 milliseconds and the electronic counter indicates 10 pulses are present during the gating time, the frequency under measurement is 2000 hertz.

SUMMARY QUESTIONS

1. Pulses of various shapes and frequencies are used in communications, control, and computer equipment. The sine-wave oscillator, which converts direct current into an alternating-current sine wave is the basic generating circuit. The sine-wave oscillator consists of a frequency-determining network; either an LC network or crystal, an amplifier to amplify the very low power output of the frequency-determining network, and a positive or in-phase feedback circuit to sustain oscillations in the frequency-determining network. In order to convert one waveshape to another, waveshaping circuits deliberately introduce distortion to the original waveshape. Examples of waveshapes used in electronic equipment are: square waves, stair-step waves, spikes, and positive and negative pulses. Terms used to describe the characteristics of pulses are: width, rise time, decay time, pulse repetition rate, and pulse repetition time.

   a. A waveshape that starts at 0 volts, rises sharply to a positive 10 volts, remains at that level, then drops sharply to 0 volts is a (an) _____ _____.

   b. A square wave consists of a sine wave and a great number of its _____ harmonics.

   c. The length of time from the start of one pulse to the start of the next similar pulse is called the _____ _____ time.

2. Limiter circuits clip or limit a portion of the signal applied to it. A diode limiter will limit or impede either the negative or positive alternation of the input signal, depending on how the diode is connected. A limiter circuit that limits a portion of both alternations of the input signal is called a squaring circuit. Two biased diodes are used, one to limit the positive alternation, the other the negative alternation. The amount of diode bias voltage determines the amplitude of the square-wave output signal.
   a. The alternation that _____-biases the diode of a diode limiter will not appear at the output.
   b. A diode squaring circuit will limit the positive and negative _____ of the input signal.
3. Differentiator and integrator circuits, consisting of resistor-capacitor circuits are used for changing pulse shapes. The output of the differentiator circuit is taken from across the resistor; the output of the integrator is taken from across the capacitor. The output-voltage waveform of the differentiator is a peaked wave or spike; its peak amplitude is equal to that of the input voltage. The output-voltage waveform of the integrator is a triangular waveform; its pulse width is twice that of the input pulse.
   a. The time constant of the RC network of an integrator is _____ compared to the input pulse width.
   b. The instantaneous voltages across the resistor and capacitor of a differentiator circuit equal the _____ voltage.
   c. The output voltage of an integrator circuit is taken across the _____; the differentiator output is taken across the _____.
4. The basic operation of a flip-flop is to provide one output pulse for every two pulses applied. The four-stage serially-connected flip-flop has the ability to provide one output pulse for every 16 input pulses. Input pulses will cause the flip-flop to alternate between its reset and its set states. Seeing that each flip-flop will divide-by-two, 16 input pulses applied to flip-flop 1 will provide eight output pulses; eight input pulses, applied to flip-flop 2 will provide four output pulses; four input pulses ap-

plied to flip-flop 3 will provide two output pulses; and two input pulses applied to flip-flop 4 will provide one output pulse. The set state of each flip-flop is represented by a logical 1, the reset state by a logical 0.
   a. A single flip-flop provides one output pulse for every _____ input pulse(s).
   b. A serially-connected four-stage flip-flop will provide one output pulse for every _____ input pulse(s).
   c. When 10 input pulses have been applied, flip-flops 1 and 3 are _____ and flip-flops 2 and 4 are _____.

5. A ring counter consists of a number of flip-flops that form a complete loop. A five-stage ring counter has all inputs applied to all flip-flops simultaneously. Each input pulse causes one flip-flop to reset and the next flip-flop in the ring to set. In this manner, only one flip-flop is in the set state at any point during the operation of the ring counter. The total number of stages required in a ring counter is determined by the number of bits, including zero, that the counter is capable of storing. For example, a ring counter capable of counting from zero through seven, requiries eight flip-flop stages.
   a. A ring counter with eight stages has _____ stage(s) in the reset state at any one point in time.
   b. The first stage of a ring counter is usually in the set state when the _____ input pulse is applied.
   c. When the last stage of a ring counter changes to the reset state, the first stage changes to the _____ state.

6. Parallel counters respond more rapidly to input pulses than cascaded or serially-connected counters. In contrast to ring counters, parallel counters require fewer flip-flop stages. A divide-by-four parallel counter requires two flip-flop stages and one AND gate. The application of four input pulses causes this counter to cycle once. The two flip-flops are initially in the reset state. After the first pulse is applied, the first flip-flop is set and the second flip-flop remains reset. After the second input pulse, the first flip-flop returns to a set state and the second flip-flop becomes set. The third input pulse sets flip-flop 1 but has no impact on flip-flop 2. After the fourth pulse, both flip-flops are again in the reset state.

a. The first flip-flop of a divide-by-four counter changes state four times after a total of _____ input pulses have been applied.
   b. The second flip-flop of a divide-by-four counter changes state _____ times after a total of four input pulses have been applied.
   c. The AND gate of a divide-by-four counter inhibits the second flip-flop from changing state when the first flip-flop is in the _____ state.
7. A divide-by-three counter completes one cycle when three input pulses are applied. The states of flip-flops can be represented by logical 1's and 0's. For example, the states of two reset flip-flops can be represented by 00. If the first flip-flop is set and second flip-flop is reset, these states can be represented as 01. Then, the sequence in which a counter changes state can be represented by logical 1's and 0's. The divide-by-three counter described in this chapter changes state as follows: 01, 11, and 00. This principle of representing the states of flip-flops by logical 1's for the set states and logical 0's for the reset states can be applied regardless of the number of stages used in the counter.
   a. The state of a counter consisting of three flip-flops can be represented by 001, which means the first stage is _____ and the second and third stages are _____.
   b. The state of a four-stage counter is represented by 1000, which means the fourth stage is _____.
   c. The state of a four-stage counter that counts in the binary number system is represented by 1000, which means _____ input pulses have been applied.
8. A divide-by-five counter completes one cycle when five input pulses have been applied. Three flip-flop stages are required, as shown in Fig. 4-18 and 4-19. Flip-flops A, B, and C are reset before the first pulse is applied. The reset state of the counter is represented by 000. After the first pulse is applied, the state of the counter is 001. The state of the counter after the second input pulse arrives is 011. The states of the counter after the third and fourth pulses are applied are 111 and 100, respectively. To reset flip-flop C, the fifth pulse must be

in coincidence with the 0 outputs from flip-flops A and B. Simultaneously, the fifth input pulse and the 1 output from flip-flop C are applied through gate Z1 so that flip-flop A remains in the reset state. Thus, after the fifth input pulse has been applied, the state of the counter is 000.

 a. Flip-flop C (Fig. 4-19) is inhibited from changing to the set state by Z4 when the second input pulse arrives because flip-flop B is in the _____ state.

 b. Flip-flop B is inhibited from changing to the set state by Z2 when the first input pulse arrives because flip-flop A is in the _____ state.

9. A decade or divide-by-ten counter produces one output pulse or cycles once for each group of ten input pulses applied. To summarize the operation of a parallel decade counter, refer to Figs. 4-20 and 4-21. Assume that five input pulses have been applied. The state of the counter at this time is 0101. This means that flip-flops A and C are set and flip-flops B and D are reset. The sixth input pulse will reset flip-flop A. However, before flip-flop A is reset, the 1 output from flip-flop A and the sixth input pulse are applied through AND gate Z1 to set flip-flop B. Flip-flop C will not change state because, in addition to the sixth input pulse, the 1 output of both flip-flops A and B must be present at the input of AND gate Z4. Flip-flop B however is reset, therefore the 1 input from flip-flop B is not present. Flip-flop D also will not change state because, in addition to the sixth input pulse, the 1 output of flip-flops A, B, and C must be present at the input of AND gate Z6. However the absence of the 1 output of flip-flop B is inhibiting AND gate Z6. The state of the counter after the sixth input pulse has been applied is 0110.

 a. AND gate Z6 (Fig. 4-20) must have the input pulse in addition to the 1 output of flip-flops A, B, and C to _____ flip-flop D.

 b. One output pulse will be obtained from a parallel decade counter after _____ input pulses have been received.

 c. The state of flip-flops D, C, B, and A after the 10th input pulse is applied will be _____.

10. Applications of counters can be grouped into two categories, direct or indirect. Direct counting is where every event or item is used in the counting process, while indirect counting is by comparison of a known frequency with the frequency to be measured. Other methods of indirect counting are frequency division and scaling. The advantages of electronic counters are reliability, or accurate counting, and high speed. One application of direct counting is the packaging of small items. A photocell senses the objects, and converts changes in light intensity to electrical pulses. The counter responds to these electrical pulses, counts to a predetermined level, then activates a mechanical device to start packaging the next unit. An application of indirect counting is the frequency meter. The unknown frequency input is applied to a counter circuit through a gate. The length of time that the gate is opened is determined by a known frequency-stable oscillator and frequency-divider circuits. By knowing how long the gate is left open, the unknown frequency can be determined.

   a. The two general categories of counter applications are _____ and _____ counting.
   b. Two advantages of counter circuits are _____ and high _____.
   c. If the frequency under measurement is 1 megahertz and the gating time is 5 milliseconds, the electronic counter would indicate that _____ pulses are present during the gating time.

SUMMARY ANSWERS

1a. A wave shape that starts at 0 volts, rises sharply to a positive 10 volts, remains at that level, then drops sharply to 0 volts is a **positive pulse.**

1b. A square wave consists of a sine wave and a great number of its **odd** harmonics.

1c. The length of time from start of one pulse to the start of the next similar pulse is called the **pulse repetition time.**

2a. The alternation that **forward**-biases the diode of a diode limiter will not appear at the output.

2b. A diode squaring circuit will limit the positive and negative **peaks** of the input signal.

3a. The time constant of the RC network of an integrator is **long** compared to the input pulse width.

3b. The instantaneous voltages across the resistor and capacitor of a differentiator circuit equal the **applied** voltage.

3c. The output voltage of an integrator circuit is taken across the **capacitor;** the differentiator output is taken across the **resistor.**

4a. A single flip-flop provides one output pulse for every **two** input pulses.

4b. A serially-connected four-stage flip-flop will provide one output pulse for every **16** input pulses.

4c. When 10 input pulses have been applied, flip-flops 1 and and 3 are **reset** and flip-flops 2 and 4 are **set.**

5a. A ring counter with eight stages has **seven** stages in the reset state at any one point in time.

5b. The first stage of a ring counter is usually in the set state when the **first** input pulse is applied.

5c. When the last stage of a ring counter changes to the reset state, the first stage change to the **set** state.

6a. The first flip-flop of a divide-by-four counter changes state four times after a total of **four** input pulses have been applied.

6b. The second flip-flop of a divide-by-four counter changes state **two** times after a total of four input pulses have been applied.

6c. The AND gate of a divide-by-four counter inhibits the

second flip-flop from changing state when the first flip-flop is in the **reset** state.

7a. The state of a counter consisting of three flip-flops can be represented by 001, which means the first stage is **set** and the second and third stages are **reset**.

7b. The state of a four-stage counter is represented by 1000, which means the fourth stage is **set**.

7c. The state of a four-stage counter that counts in the binary number system is represented by 1000, which means **eight** input pulses have been applied.

8a. Flip-flop C (Fig. 4-19) is inhibited from changing to the set state by Z4 when the second input pulse arrives because flip-flop B is in the **reset** state.

8b. Flip-flop B is inhibited from changing to the set state by Z2 when the first input pulse arrives because flip-flop A is in the **reset** state.

9a. AND gate Z6 (Fig. 4-20) must have the input pulse in addition to the 1 output of flip-flops A, B, and C to **set** flip-flop D.

9b. One output pulse will be obtained from a parallel decade counter after **10** input pulses have been received.

9c. The state of flip-flops D, C, B, and A after the 10th input pulse is applied will be **0 0 0 0**.

10a. The two general categories of counter applications are **direct** and **indirect** counting.

10b. Two advantages of counter circuits are **reliability** and high **speed**.

10c. If the frequency under measurement is 1 megahertz and the gating time is 5 milliseconds, the electronic counter would indicate that **200** pulses are present during the gating time.

5

Special Semiconductor Devices

What You Will Learn
The many special semiconductors that have been developed in recent years will be discussed in this chapter. Each device will be discussed in terms of its structure, operating characteristics, and circuit applications. Space, however, does not permit discussion of all of the applications of each semiconductor device. The special semiconductors which will be analyzed are zener diodes, the varistor, photoresistors and photodiodes, unijunction transistors, silicon controlled rectifiers, tunnel diodes, and PNPN transistors. (See Fig. 5-1).

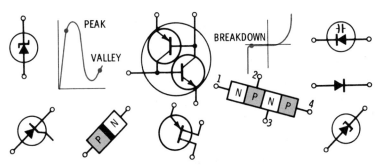

Fig. 5-1. Semiconductor devices.

ZENER DIODE

The zener diode is one of the semiconductor devices which, due to its chemical construction, reacts in a manner unlike the normal semiconductor diode. In this concept, we will discuss the characteristics and operating principles of the zener diode.

Semiconductor Theory

Fig. 5-2 shows the direction of electron flow through the P-N junction when the junction is forward-biased. The voltage-current characteristic curve, Fig. 5-3, shows that the

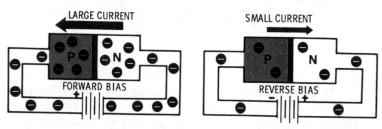

Fig. 5-2. P-N junction biasing.

higher the forward-bias voltage, the greater the electron flow through the P-N junction. Fig. 5-2 also shows the direction in which electrons will try to flow through the P-N junction when it is reverse-biased. It should be noted in Fig. 5-3, that when the junction is reverse-biased, current, for all practical purposes, is zero.

Zener Diode P-N Junction

The zener diode differs from the conventional semiconductor diode in that the amount of impurity elements added to the pure germanium or silicon is very closely controlled. By controlling the amount of impurity elements added, a con-

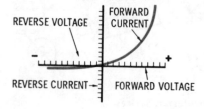

Fig. 5-3. P-N junction voltage-current characteristic curve.

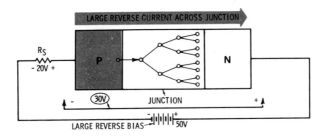

Fig. 5-4. Reverse-biased zener diode.

dition called *electron multiplication by collision* is created when the P-N junction is reverse-biased to a critical value. As shown in Fig. 5-4, when sufficient reverse-bias voltage is applied, a free electron from the P-type element will collide with a fixed electron and knock it free. These two electrons will, in turn, collide again with other fixed electrons, knocking two more free. This electron collision will continue, each time multiplying the number of electrons freed. This current multiplication process increases rapidly at the critical reverse-voltage level, as shown in Fig. 5-5, causing a constant voltage to appear across the zener diode,

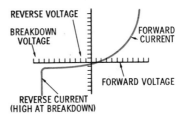

Fig. 5-5. Zener diode voltage-current characteristic curve.

a high reverse current to pass through it, and the remaining voltage to be dropped across series resistor R_s. The reverse-bias voltage level at which this high reverse current passes is called the *breakdown voltage* of the zener diode.

Q5-1. Electron multiplication in zener diodes is obtained by controlling the amount of _____ added.

Q5-2. When breakdown voltage is applied to a zener diode reverse current is _____.

Q5-3. When breakdown voltage is applied to a zener diode forward current is _____.

167

> **Your Answers Should Be:**
> **A5-1.** Electron multiplication in zener diodes is obtained by controlling the amount of **impurity elements** added.
> **A5-2.** When breakdown voltage is applied to a zener diode reverse current is **high**.
> **A5-3.** When breakdown voltage is applied to a zener diode forward current is **zero**.

ZENER DIODE APPLICATIONS

The primary function of the zener diode is voltage regulation. The zener diode is sensitive to voltage changes above predetermined levels, just as a fuse is sensitive to current changes. Due to breakdown characteristics in its reverse-bias condition, the zener diode can effectively prevent variations from a preselected voltage value.

Zener Diode Symbol

As shown in Fig. 5-6, two symbols are used for the zener diode; one has the cathode in the form of a "Z," the other has the current breakdown curve as part of the symbol.

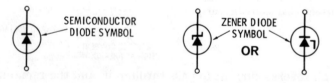

Fig. 5-6. Semiconductor diode and zener diode symbols.

Zener Diode Clipper Circuit

A matched pair of zener diodes can be used to provide a square-wave output from a sine-wave input. As shown in Fig. 5-7, the breakdown rating of the two matched zener diodes will determine the amplitude of the square-wave output voltage. When one zener diode is reverse-biased by the a-c input voltage, the other is forward-biased. During the opposite half-cycle of the a-c input voltage, the reverse/for-

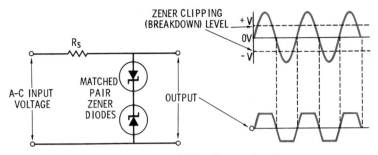

Fig. 5-7. Zener diode clipper circuit.

ward-bias condition switches, producing a square-wave output voltage whose amplitude is equal to the breakdown level.

Zener Diode Meter-Protection Circuit

A zener diode can be used as a protective device for meter movements. The zener diode has the advantage over a fuse in this application in that it will function indefinitely without replacement. As shown in Fig. 5-8, a zener diode with a breakdown-voltage value equal to the applied-voltage value which will cause full-scale meter deflection of the meter, is placed in parallel with the meter. When a voltage in excess of the meter full-scale–deflection value is applied, the zener diode will break down, shunting the excess current away from the meter.

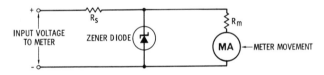

Fig. 5-8. Zener diode meter-protection circuit.

Q5-4. The zener diode schematic symbol has two legs forming a Z on the _____ element.

Q5-5. The zener diode differs from a fuse in that in its reverse-bias direction is sensitive to _____ overloads.

Q5-6. If the zener diodes used in a clipper circuit have a breakdown rating of 12 volts, the amplitude of the output square wave will be _____ volts.

Your Answers Should Be:

A5-4. The zener diode schematic symbol has two legs forming a Z on the **cathode** element.

A5-5. The zener diode differs from a fuse in that in its reverse-bias direction it is sensitive to **voltage** overloads.

A5-6. If the zener diodes used in a clipper circuit have a breakdown rating of 12 volts, the amplitude of the output square wave will be **12** volts.

Zener Diode Arc Suppression

Zener diodes are used as surge protection devices across inductive loads where "kickback" voltages cause arcing at switch contacts and noise pickup in adjacent equipment.

As shown in Fig. 5-9, when the switch is opened the voltage induced in the relay coil by its collapsing field is often very much greater than the normal applied voltage. This very high "kickback" voltage developed across the relay coil can cause arcing across the switch contacts. By using a zener diode that has a breakdown voltage slightly higher than the applied voltage, the kickback voltage can be reduced.

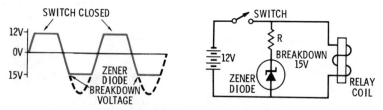

Fig. 5-9. Zener diode arc-suppression circuit.

When the kickback voltage reverse-biases the zener diode to its breakdown level, the very high kickback voltage is reduced to the breakdown level of the zener diode.

Zener Diode Voltage Regulation

The zener diode, when breakdown voltage is applied, will have a very high reverse current through it, but it will have a constant voltage drop across its P-N junction. This constant-voltage value is the value at which the zener diode will provide voltage regulation. Zener diodes are available with breakdown-voltage ratings, or voltage-regulation ratings, of between 2 and several hundred volts.

The zener diode shown in Fig. 5-10 has a breakdown voltage of 100 volts. That is, if any voltage in excess of 100 volts is applied to it, the zener diode will break down, causing a high reverse current through the zener diode, and a constant

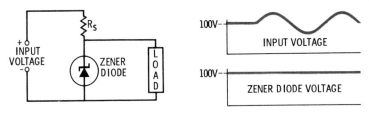

Fig. 5-10. Zener diode voltage regulation.

100-volt drop across it; all voltage in excess of 100 volts is dropped across series resistor R_s. Consequently, the voltage applied to the load will have a constant value of 100 volts, even though the input voltage increases to above 100 volts.

A-C Voltage Regulation

Power supplies with multisecondary winding transformers can be regulated by using a matched pair of 58-volt zener diodes across the primary winding, as shown in Fig. 5-11. As the primary voltage fluctuates above 115 volts alternating current, the zener diodes alternately breakdown, providing a regulated a-c input voltage to the power supply.

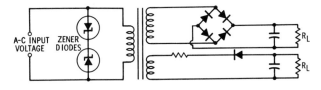

Fig. 5-11. Zener diode a-c voltage regulation.

Q5-7. A zener diode used for arc suppression must have a breakdown voltage _____ than the applied d-c voltage.

Q5-8. The voltage drop across a zener diode is constant at the diode's rated _____ voltage.

Your Answers Should Be:

A5-7. A zener diode used for arc suppression must have a breakdown voltage **higher** than the applied d-c voltage.

A5-8. The voltage drop across a zener diode is constant at the diode's rated **breakdown** voltage.

SPECIAL PURPOSE SEMICONDUCTORS

Semiconductors, due to their chemical construction and physical size, have been utilized for many applications. The semiconductors that we will discuss here are but a few of the many special devices that have developed through man's knowledge.

Semiconductor Used as a Variable Capacitor

A semiconductor diode can be used as a variable capacitor, or varactor, by varying the reverse-bias voltage across it. A typical varactor will change capacitance from 200 picofarads when reverse-biased with 1 volt, to 50 picofarads when reverse-biased with 12 volts. Raising the reverse-bias voltage lowers the capacitance, and lowering the reverse-bias voltage raises the capacitance. Fig. 5-12 shows how the varactor changes capacitance as the reverse-bias voltage is changed. When the varactor reverse bias is low, the majority charges move a small distance away from the P-N junction, but when a high reverse bias is applied, the majority

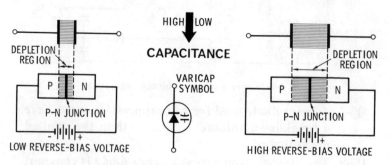

Fig. 5-12. Semiconductor variable capacitor (varactor).

charges move a greater distance away from the P-N junction. This *depletion region*, or the area of separation between charges in the P-type and N-type semiconductor materials, determines the capacitance value of the varactor. The depletion area is like the dielectric area between the plates of a conventional capacitor. Applications for the varactor include uses in amplitude and frequency modulators, afc circuits, and variable-frequency oscillator circuits.

Semiconductor Photoresistor

A *photoresistor*, or a resistor which changes its resistance with a change in light intensity, is made from a single piece of pure germanium or silicon—N-type or P-type semiconductor material. Fig. 5-13 shows that raising the intensity of the light focused on the photoresistor frees more electrons in the semiconductor material, allowing the semiconductor to conduct more (lower resistance); decreasing the light intensity produces the opposite effect. Pure semiconductors are used for high-resistance photoresistors and either N types or P types are used for low-resistance photoresistors. Some of the typical applications are in photographic light meters and automatic lens-adjusting cameras.

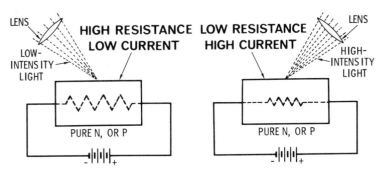

Fig. 5-13. Photoresistor.

Q5-9. The capacitance of a varactor is determined by the area of the _____ _____.

Q5-10. If the light intensity applied to a photoresistor is increased, its resistance will _____.

Q5-11. Raising the reverse-bias voltage of a varactor will _____ the capacitance value.

Your Answers Should Be:

A5-9. The capacitance of a varactor is determined by the area of the **depletion region**.

A5-10. If the light intensity applied to a photoresistor is increased, its resistance will **decrease**.

A5-11. Raising the reverse-bias voltage of a varactor will **decrease** the capacitance value.

Semiconductor Photodiode

The *photodiode* is a special P-N junction diode made with an opening in the diode case to focus light on the P-N junction. Light focused on the junction raises the number of unlike charges at the junction which, in turn, raises the conductivity of the diode. As shown in Fig. 5-14, the diode is

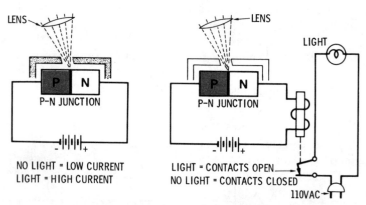

Fig. 5-14. Photodiode and automatic light control.

reverse-biased so that there is very little current when it is in the dark. Exposing it to light makes it conduct more.

Photodiode Automobile Light-Dimmer Circuit

The photodiode automobile light-dimmer circuit shown in Fig. 5-15 will automatically change headlights from high to low when the photodiode is activated by light. The photodiode is reverse-biased by the battery voltage when the high-beam switch is depressed. The reverse-biased photodiode will not allow current through the relay coil due to its extremely high resistance. Relay contacts A and C complete the battery circuit to the high-beam headlights. When oncoming headlights focus light on the photodiode P-N junction, the conductivity of the diode increases, overcoming the reverse bias, allowing current through the relay coil. Energized relay contacts A and B complete the battery circuit to the low-beam headlights. When the light source is re-

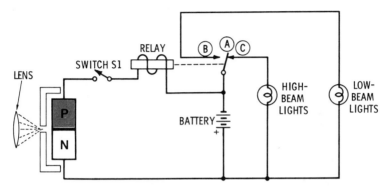

Fig. 5-15. Photodiode automobile light-dimmer circuit.

moved from the photodiode, its resistance again increases, disabling the relay, and turning on the high beams.

Photovoltaic Cells

The photovoltaic cell, or solar cell as it is more commonly called, is similar in construction to the photodiode, but instead of applying a voltage to it, it is used to generate a voltage by focusing light on the P-N junction. When the solar cell is exposed to bright light it will generate about 0.4 volt. The most common solar cell, about 1/2-inch square, will generate 0.5 watt of power. Combining cells in series-parallel produces higher voltage and current capacities.

Q5-12. **When light is focused on the P-N junction of a photodiode its resistance** _____.

Q5-13. **In order for current to pass through a photodiode, the light focused on its P-N junction must overcome its** _____ _____.

Q5-14. **Five photovoltaic cells connected in series will produce a voltage of** _____ **volts.**

> **Yours Answers Should Be:**
>
> A5-12. When light is focused on the P-N junction of a photodiode its resistance **decreases**.
>
> A5-13. In order for current to pass through a photodiode, the light focused on its P-N junctions must overcome its **reverse bias**.
>
> A5-14. Five photovoltaic cells connected in series will produce a voltage of **2 volts**.

SILICON CONTROLLED RECTIFIER

The silicon controlled rectifier (scr), as its name implies, is a semiconductor device whose basic material is silicon. It is very similar to a silicon diode. The major difference is that it is a *controlled rectifier;* that is, an additional element called a *gate* is added to control the period of time that the scr will act as a diode.

SCR Symbol and Operation

Fig. 5-16 shows an scr and its schematic symbol. The heavy bottom contact is the anode, the larger of the two top leads is the cathode, and the small top lead is the gate. The anode-to-cathode circuit, when no gate voltage is applied, is identical to the conventional semiconductor diode. It is non-conducting and has very high resistance in both the forward and reverse directions. The gate-to-cathode circuit is identical to a small diode; that is, when the gate voltage is applied, the scr resembles the conventional semiconductor diode.

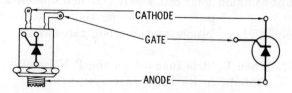

Fig. 5-16. Silicon controlled rectifier symbol.

The amount of gate voltage required to switch the forward resistance of the scr from high to low is very small. Fig. 5-17 shows the relationship between gate and anode-to-cathode voltage and cathode-to-anode current. During Time 1, the gate voltage is applied and the cathode is positive, therefore, the scr has a high resistance in both directions and there is no forward current. During Time 2, when the scr anode is positive, the gate voltage is applied, and the forward resistance decreases greatly, allowing forward cur-

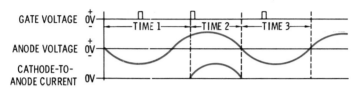

Fig. 5-17. Silicon controlled rectifier characteristic curve.

rent. Even though the gate voltage is removed, forward current will continue as long as the scr anode is positive. During Time 3, gate voltage is applied, but the scr anode is negative, therefore the resistance remains high and there is no forward or reverse current.

SCR Application

One of the major applications of the scr is that of a switching control. Fig. 5-18 shows two scr's used as an a-c control circuit. When a small gate signal is applied to the scr's the circuit acts as a closed switch, but when the gate signal is removed the circuit acts as an open switch.

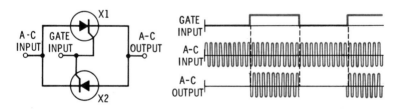

Fig. 5-18. Scr a-c control circuit.

Q5-15. The three elements of the scr are the _____, _____, and _____.

Q5-16. A conducting scr will continue to conduct until the positive _____ voltage is removed.

> **Your Answers Should Be:**
> **A5-15.** The three elements of the scr are the **anode, cathode,** and **gate.**
> **A5-16.** A conducting scr will continue to conduct until the positive **anode** voltage is removed.

UNIJUNCTION TRANSISTOR

The unijunction transistor (ujt) was originally called a *double-base diode* because it is similar to the conventional diode, except that it has two base connections instead of one.

Construction and Symbol

The ujt consists of a block of N-type material with a connection at each end, and a P-type material at the center of

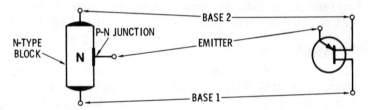

Fig. 5-19. Unijunction transistor construction and symbol.

the block, forming a typical P-N junction. As shown in Fig. 5-19, the two end connections are called *base 1* and *base 2*, while the P-N junction connection is called the *emitter*.

Operation

The ujt and equivalent circuit of Fig. 5-20 show that the ujt can be considered as two series resistors center-tapped by a semiconductor diode. This illustration is used to explain the basic ujt operation, rather than a practical circuit.

When a voltage is applied between base 1 (B1), and base 2 (B2) it produces a uniform voltage drop across the internal resistance of the N-type block. For this discussion, a voltage of 20 volts will be used; however, ujt's operate with voltages between 3 and 40 volts or more. Since the emitter is in the center of the bar, the voltage drop between the emitter and D_1 is +10 volts reverse-biasing the emitter-to-B1 junction. If a positive voltage in excess of 10 volts is applied to the emitter-to-B1 input terminals, the junction is forward-biased, causing B1-to-emitter current. As there is

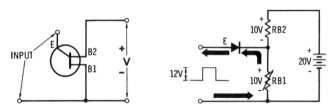

Fig. 5-20. Unijunction transistor circuit and equivalent circuit.

current, resistance R_{B1} (B1-to-emitter) drops sharply to a very low value. Emitter-B1 input voltage switches the ujt from reverse to forward bias, causing a sharp rise in B1-to-emitter current and an equally sharp decrease in resistance.

Unijunction Transistor Characteristic Curve

The characteristic curve of Fig. 5-21 shows that the ujt has three operational regions: cutoff, peak point, and valley point. The cutoff region is where the emitter is reverse-biased. As forward-bias voltage fires the ujt, emitter current increases sharply and emitter-B1 voltage drops sharply. This occurs at the peak point. At the valley point the emitter-B1 voltage is the minimum for a positive emitter current.

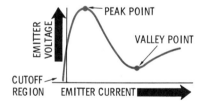

Fig. 5-21. Unijunction transistor characteristic curve.

Q5-17. The ujt has an emitter element and two base connections called _____ and _____.

Q5-18. The ujt will conduct only when a positive voltage overcomes the emitter-to-B1 _____ bias.

Q5-19. Three important operating regions of the ujt are: _____, _____ _____ and _____ _____.

> **Your Answers Should Be:**
> **A5-17.** The ujt has an emitter element and two base connections called **base 1** and **base 2**.
> **A5-18.** The ujt will conduct only when a positive voltage overcomes the emitter-to-B1 **reverse** bias.
> **A5-19.** The three important operating regions of the ujt are: **cutoff, peak point,** and **valley point.**

UJT CIRCUIT APPLICATIONS

The ujt, because of its unique characteristic curve, is ideally suited for use in timing circuits and pulse-generating and sawtooth-wave generating circuits.

UJT Sawtooth-Oscillator Circuit

A typical ujt sawtooth-oscillator circuit is shown in Fig. 5-22. Capacitor C1 will charge through R1 until the value of the voltage developed across C1 is sufficient to forward-bias the B1-to-emitter junction. As soon as the voltage reaches the forward-bias value, the ujt fires, causing a very large B1-to-emitter current. As this rapid change from *cutoff* to

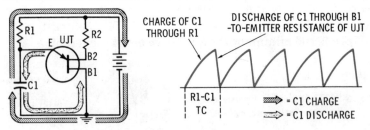

Fig. 5-22. Ujt sawtooth oscillator circuit and waveforms.

peak current occurs, the B1-to-emitter resistance drops sharply, causing C1 to discharge quickly through this low resistance. When C1 discharges, current reaches the *valley point*. The voltage across C1 is insufficient to keep the ujt forward-biased, so it returns to its nonconducting state. This cycle will continue to repeat itself. The frequency of the output sawtooth voltage is determined primarily by the RC time constant of R1-C1. The amplitude of the sawtooth voltage is determined primarily by the applied voltage and the emitter *firing* voltage characteristics of the ujt.

UJT Multivibrator Circuit

In the multivibrator circuit shown in Fig. 5-23, when plus voltage is applied, C1 will charge through R2 and forward-biased diode X1. The ujt is cut off until the charge voltage across C1 equals the value necessary to forward-bias the emitter-to-B1 junction of the ujt. At this instant, the ujt *fires*, causing the voltage at the ujt emitter to drop to nearly 0 volts. This sharp drop in voltage reverse-biases diode X1. With X1 cut off, the ujt B1-to-emitter current must now pass through R2. At the same time, C1 is discharging through R1, until the cathode of X1 is at the same potential as the emitter of the ujt. At this time, X1 becomes forward-biased and starts to conduct again. As X1 conducts, B1-to-emitter current drops to the valley-point, causing the ujt to cut off. Capacitor C1 starts to charge and the cycle repeats.

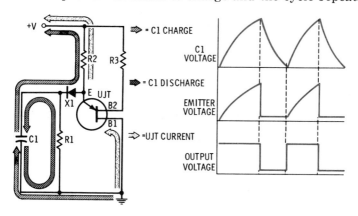

Fig. 5-23. Ujt multivibrator circuit and waveforms.

Q5-20. When the ujt B1-to-emitter current is at its peak value, B1-to-emitter resistance is _____.

Q5-21. The ujt sawtooth-oscillator–output frequency is determined by ujt circuit _____ _____ _____.

Q5-22. The ujt multivibrator will remain cut off until the voltage across _____ equals the ujt firing voltage.

> **Your Answers Should Be:**
> **A5-20.** When the ujt B1-to-emitter current is at its peak value, B1-to-emitter resistance is **low**.
> **A5-21.** The ujt sawtooth-oscillator–output frequency is determined by the ujt circuit **RC time constant**.
> **A5-22.** The ujt multivibrator will remain cut off until the voltage across **C1** equals the ujt firing voltage.

TUNNEL DIODE CHARACTERISTICS

The tunnel diode is another unique device that, since its development, has found many uses in high-frequency switching mechanisms and flip-flop storage devices.

The tunnel diode (td) schematic symbols are shown in Fig. 5-24. The anode or plate is in the shape of its characteristic curve or the cathode can be distinguished by its semicircular shape instead of the conventional triangle.

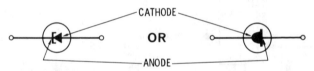

Fig. 5-24. Tunnel diode symbols.

Tunnel Diode Operating Characteristics

The td switches between two stable states: peak point and valley point. Fig. 5-25 shows that the value of load resistance R_L provides a plate current (A) less than the peak-current value. This value of plate current produces a minimum output-voltage condition. With no input-trigger pulse applied, the td remains in this high-current state.

When a short-time-duration positive trigger (turn-on) pulse is applied, both voltage and current of the td increase to peak point, (B) in Fig. 5-25. When the turn-on pulse is removed, the circuit *switches* to a point just to the right of the valley, point (C) in Fig. 5-25.

As the current starts to decrease from the peak point, the

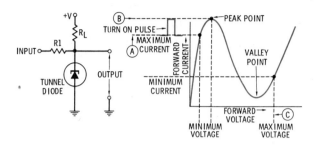

Fig. 5-25. Tunnel diode characteristic curve.

voltage across the td increases. This increased voltage drop across the td decreases the current through the series load resistor and the td. Again, less current through the td increases the voltage drop across it. This process continues until the valley point is reached. At the valley point, the forward-bias voltage across the td causes it to act as a normal diode. The current through the td and the series-connected load resistor causes a stable operating point, shown at point (C) in Fig. 5-25. The circuit remains in this stable state until a negative trigger pulse is applied.

The area between operating points B and C of Fig. 5-25 is called the *negative resistance* region. *A decrease in current through the diode produces an increased voltage drop across it.* This negative resistance characteristic makes the tunnel diode different from the conventional diode.

When a short-time-duration negative turn-off pulse is applied, forward-bias voltage of the td decreases. This decrease in td forward-bias voltage causes its current to drop to the valley point, which lowers the voltage drop across the td. As the voltage drops, more current passes through the series load resistor, causing a further drop in td voltage. This process continues until the minimum output point (A, Fig. 5-25) is reached. The circuit remains in this stable state until the application of the next turn-on pulse.

Q5-23. When a positive trigger pulse is applied to the td it will increase to its _____ _____ value.

Q5-24. Minimum td current is called the _____ _____ value.

183

> **Your Answers Should Be:**
>
> A5-23. When a positive trigger pulse is applied to the td it will increase to its **peak point** value.
>
> A5-24. Minimum TD current is called the **valley point** value.

TUNNEL DIODE AND GATE AND OR GATE

The tunnel diode can perform the AND and OR logic functions with a minimum of circuit components, high speed, low power consumption, and very stable operating characteristics.

Tunnel Diode Switching Time

The td is the fastest switching device known to date. Fig. 5-26 shows that the typical time required to switch from one steady-state condition to the other is about 27 picoseconds (27×10^{-12} seconds), or the time that it takes light to travel 0.3 inch. Therefore, the td is ideally suited for high-speed switching circuits.

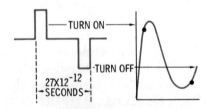

Fig. 5-26. Tunnel diode switching time.

Tunnel Diode OR Gate

The OR gate, shown in Fig. 5-27, requires either input A OR input B to switch the td from its low voltage-high current steady state to its high voltage-low current state.

Notice that because of the high input resistance, the amplitude of either input A or B is sufficient to increase the forward bias to the value required to increase td current to its peak point to start switching it to its opposite steady state. An external negative turn-off pulse must be applied to return the td to the original low voltage-high current steady state.

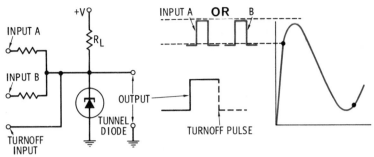

Fig. 5-27. Tunnel diode OR gate.

Tunnel Diode AND Gate

The AND gate shown in Fig. 5-28 requires both inputs A AND B to switch the td from its low voltage-high current steady state to the reverse steady state. Notice that because of the low input resistance, the amplitude of input A alone

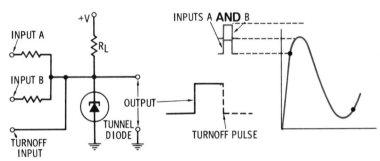

Fig. 5-28. Tunnel diode AND gate.

is not sufficient to increase td current to its peak point; both inputs A and B are required. As with the OR gate, a negative turn-off pulse is required to return to the original steady state.

Q5-25. The tunnel diode is ideally suited for logic circuits because of its rapid _____ time.

Q5-26. Both td AND and OR gates require a (an) _____ pulse to return the td to its original state.

185

> **Your Answers Should Be:**
> A5-25. The tunnel diode is ideally suited for logic circuits because of its rapid **switching** time.
> A5-26. Both td AND and OR gates require a **turn-off** pulse to return the td to its original state.

PNPN DEVICES

The PNPN device, because of its structure, has countless applications and comes in several different forms. In this concept we will discuss the various configurations of this family of PNPN devices.

The PNPN device, as its name indicates, is constructed of four layers of semiconductor material, usually silicon. The

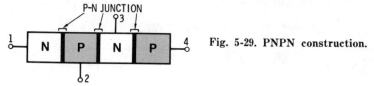

Fig. 5-29. PNPN construction.

layers are alternately doped with P- and N-type inpurities, as shown in Fig. 5-29. These four layers of N-type and P-type material form three PN junctions.

Symbolic Representations

Fig. 5-30 shows that PNPN devices have several schematic symbols. The difference between the schematic symbols is that all use the outside layers, while some have connections to either one or both inner layers.

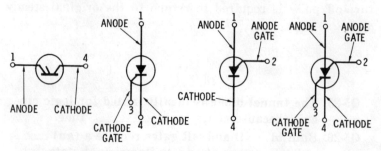

Fig. 5-30. PNPN symbolic representations.

Operation

Fig. 5-31 shows how the PNPN device is biased and the interrelationship between all regions of the device. The PNPN device is drawn as two interconnected transistors, one an NPN, the other a PNP. When properly biased, the center P-N junctions are reverse-biased, and the two outside P-N junctions are forward-biased. Comparing this to conventional transistors, the emitter-base junction is always forward-biased, and the collector-base junction is always

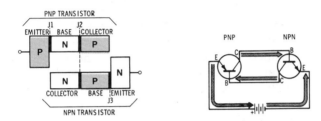

Fig. 5-31. PNPN device shown as NPN-PNP connected transistor.

reverse-biased. The P-N junction of the PNP transistor connected to the positive side of the battery is forward-biased and therefore is the emitter-base junction. The remaining P-N junction of the PNP transistor is reverse-biased and is the collector base junction. The P-N junction of the NPN transistor connected to the negative side of the battery is forward-biased, and therefore is the emitter-base junction. The remaining P-N junction of the NPN transistor is reverse-biased and is the collector-base junction. The base of the PNP transistor is connected directly to the collector of the NPN transistor. The collector of the PNP transistor is connected directly to the base of the NPN transistor.

Q5-27. The PNPN device consists of _____ layers of semiconductor material.

Q5-28. The PNPN device can be considered as two interconnected transistors, one _____, the other _____.

Q5-29. The conventional transistor emitter-base junction is always _____-biased.

Your Answers Should Be:

A5-27. The PNPN device consists of **four** layers of semiconductor material.

A5-28. The PNPN device can be considered as two interconnected transistors, one **PNP**, the other **NPN**.

A5-29. The conventional transistor emitter-base junction is always **forward**-biased.

Characteristic Curve

The voltage-current characteristic curve of the PNPN is shown in Fig. 5-32. When the device is reverse-biased, the two outer junctions are reverse-biased, and there is practically no reverse current until the reverse-bias voltage reaches a value large enough to cause breakdown. The value of reverse-bias voltage required to cause breakdown is very high and consequently has no special significance in the operation of the PNPN device.

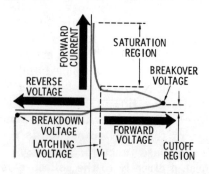

Fig. 5-32. PNPN voltage-current characteristic curve.

When forward voltage is applied, there is only a very small amount of current until the forward voltage is increased to a value sufficient to cause *breakover* of the PNPN. This small amount of current is so insignificant that the device is considered to be cut off. The breakover voltage is the value which will cause an abrupt change from a very small forward current (cutoff) to saturation current (maximum current).

If the forward voltage is now reduced, the device will remain ON (conducting) until the current has decreased to the lower extreme of the saturation region. This value of current and the corresponding voltage is called the *latching voltage*, or the minimum value of voltage that will hold the device ON. If forward voltage drops below this critical value the device will switch to its cut-off state.

The Silicon Controlled Switch

The simple silicon controlled switch (scs), shown in Fig. 5-33A, uses only the two outside connections of the PNPN device. The gated scs, on the other hand, has connections to all four elements. As shown in Fig. 5-33B the gated scs inner elements are the *cathode gate* and the *anode gate*. Where

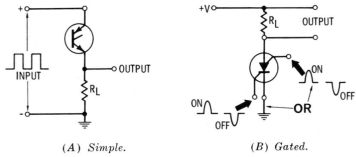

(A) *Simple.*   (B) *Gated.*

Fig. 5-33. Silicon controlled switch circuit applications.

only one or the other inner element has an external connection, it is referred to as the gate of the outside element to which it is the closest. As an example, if an external connection is made to the inner element closest to the cathode, the gate is called the *cathode gate*.

Some of the major applications of the scs are in switching circuits. The gated scs has the major applications because of the ability to apply either positive or negative gating pulses of very low voltage to switch states.

Q5-30. The voltage value required to cause switchover from cutoff to saturation is the _____ voltage.

Q5-31. The minimum voltage value which will hold current in the saturation region is the _____ voltage.

Q5-32. The major application of the scs is in low-voltage, low-current _____ circuits.

> **Your Answers Should Be:**
> **A5-30.** The voltage value required to cause switchover from cutoff to saturation is the **breakover** voltage.
> **A5-31.** The minimum voltage value which will hold current in the saturation region is the **latching** voltage.
> **A5-32.** The major application of the scs is in low-voltage, low-current **switching** circuits.

SUMMARY QUESTIONS

1. The zener diode, unlike the conventional diode, provides maximum current when the P-N junction is reverse-biased. By closely controlling the quantity of impurity materials added to the basic semiconductor material, the zener diode formed reacts to a critical value of reverse-bias voltage. The applied reverse bias causes a condition within the molecular structure of the diode that is called electron multiplication by collision. The reverse bias frees electrons from the P material that bombard and collide with other free electrons. This electron collision process continues and results in a high reverse current through the diode. Forward current is practically zero through the zener diode during the time reverse-bias voltage is applied.
   a. The quantity of impurities added to a zener diode determines the critical value of applied _____ bias to which the diode reacts.
   b. Reverse current through a zener diode is high when the proper _____ voltage is applied.
   c. When reverse current through a zener diode is high, the forward current is _____.
2. Symbols for the zener diode and conventional diode are shown in Fig. 5-6. Principal applications for zener diodes are voltage regulation, clippers or amplitude limiters, and over-voltage protective devices. Fig. 5-7 illustrates how a sine-wave input voltage is converted to a square-wave output voltage by a zener diode clipper cir-

cuit. Fig. 5-8 shows the versatile zener diode in use as an over-voltage protective device for a meter movement. The zener diode passes high reverse current when a voltage in excess of some value determined by diode manufacturing design is applied to the circuit.
  a. If the breakdown voltages of the diodes shown in Fig. 5-7 were not equal, the _____ of the output-voltage waveform would be affected.
  b. If the voltage applied to the circuit shown in Fig. 5-8 were reversed, high _____ current through the zener diode would divert current from the meter circuit.
3. Collapsing magnetic fields cause arcing at switching contacts and generate noise which can disturb the operation of electronic transmission and communication equipment in the immediate vicinity. Fig. 5-9 shows a zener diode in a circuit that suppresses switch arcing and minimizes noise generation. When the kickback voltage from the relay coil reaches the level of the diode-breakdown voltage, the amplitude is limited by the reverse-biased diode. With the breakdown voltage applied, the zener diode provides an alternate path for current. Fig. 5-10 shows how a zener diode regulates voltage. The diode can be considered as a variable current device that provides a constant output voltage. Any input voltage in excess of the diode-breakdown voltage is dropped across series resistor R_s. Fig. 5-11 shows two zener diodes connected back-to-back to limit the amplitude of an a-c input voltage. When the input voltage rises above a predetermined value, each diode alternately breaks down and a regulated a-c voltage is applied to the primary of the transformer.
  a. The direction of current when the switch shown in Fig. 5-9 is opened is _____ _____ the current supplied by the battery.
  b. If the input voltage shown in Fig. 5-10 is reversed the voltage across the load would be approximately _____.
4. A varactor is a semiconductor device that acts as a variable capacitor. The capacitance of the device is dependent on the applied voltage. As shown in Fig. 5-12, the

191

depletion area of a varactor is affected by the magnitude of applied reverse bias. When the reverse-bias voltage is increased, the capacitance of the varactor decreases. With decreased reverse-bias voltage, the capacitance of the varactor increases. Another semiconductor device is the photoresistor, which acts as a variable resistor with change in light intensity focused on the device. Increasing the intensity of light causes the resistivity of the photoresistor to decrease. Conversely, decreasing the intensity of light focused on the photoresistor causes the resistivity to increase.

   a. The capacitance of a varactor is dependent on the magnitude of _____-_____ voltage.

   b. When the depletion area of a varactor is increased, the capacitance is _____.

   c. When the intensity of light focused on a photoresistor is increased, the current through the associated circuit is _____ .

5. The photodiode is another semiconductor device that is finding wide application. The photodiode is essentially an on-off switch that is sensitive to light. With little or no light applied to the P-N junction, the device acts as an open switch. When light of a predetermined intensity is applied to the photodiode, it acts as a closed switch. Another semiconductor device is the photovoltaic cell, or solar cell. Like the photodiode, the solar cell is light sensitive. However, the solar cell is used to convert light energy to electrical energy.

   a. If the source of light is kept from reaching the P-N junction shown in Fig. 5-14, the relay contacts will _____.

   b. With switch S1 closed, the photodiode shown in Fig. 5-15 can be excluded as a source of trouble by shorting the photodiode leads and observing that the high-beam lights are switched _____.

   c. Four solar cells, of the type described, when connected in series-parallel will supply _____ volt(s), providing a source of light is available.

6. The silicon controlled rectifier (scr) is a gated or controlled diode. The addition of a gate element determines when the scr will conduct. When the scr is forward-

biased and a very low pulse voltage is applied to the gate, the scr conducts. After the pulse voltage has been removed, the scr will continue to conduct as long as the anode is positive. To cut off the scr, the anode voltage must be reduced to zero. One of the major applications of the scr is as a high-voltage/high-current a-c switching device. The advantage of the scr is that a very low pulse voltage can be used to control an extremely high voltage.
  a. The resistance of the scr is _____ until a gate pulse is applied.
  b. The scr is used as a switching device in _____ _____ circuits.
  c. With a gate pulse applied, the cathode must be _____ to cause the scr to conduct.
7. The unijunction transistor (ujt) or double-base diode consists of a block of N-type silicon or germanium with a P-type material at the center of the block. The point of contact between the P-type material and the N-type block forms a PN junction. The three external connections of the ujt are called base 1, base 2, and emitter. When voltage is supplied between the two base connections, the base 1-to-emitter junction is reverse-biased and there is no current. When a positive pulse in excess of the reverse-bias voltage is applied between emitter and base 1 the ujt will switch from cutoff to its peak-current value. When the positive input pulse is removed, ujt current will drop to its valley-point value.
  a. The three elements of the ujt are _____, _____, and _____.
  b. When a positive pulse voltage is applied between emitter and base 1 of the ujt, the emitter-base 1 junction will become _____-biased.
  c. The operational region of peak point current will be reached only when the emitter-to-base 1 junction is _____-biased.
8. The ujt, because of its operating characteristics, is used primarily in waveform-generating circuits. The ujt sawtooth oscillator shown in Fig. 5-22 develops a sawtooth-voltage output waveform across the emitter-to-base capacitor. When voltage is applied between base 1 and

base 2, the capacitor will charge to a value which will exceed the reverse bias of the emitter-to-base 1 junction, causing the ujt to conduct. The low resistance of the conducting ujt will rapidly discharge the capacitor, producing the sawtooth output. The sawtooth output frequency is determined by the time constant of the RC emitter-base 1 circuit. The ujt multivibrator shown in Fig. 5-23 operates in basically the same manner as the sawtooth generator. The addition of a diode and resistor in the emitter circuit is the only difference. The diode is reverse-biased as the ujt conducts, causing the capacitor to discharge through the added resistor, therefore producing a square-wave output.

   a. The time constant of the resistor and capacitor in the ujt emitter-base 1 circuit determines sawtooth-voltage _____.

   b. The emitter-to-base 1 resistance of the ujt is low when base 1-to-emitter current is at its _____ value.

   c. The emitter diode of a ujt multivibrator is reverse-biased while the ujt is _____.

9. The tunnel diode (td) is a two-element device that is used primarily in logic-circuit applications because of its extremely rapid switching time. It can be identified and distinguished from the conventional diode on schematics because of its characteristic-curve–shaped cathode. When input voltage is applied to the td it will switch between one of two stable states; peak-point and valley-point current. During this transition, the td exhibits a characteristic called negative resistance; a decrease in current through the td increases the voltage drop across it. A positive input pulse is required to switch td current from peak-to-valley current points; a negative input pulse is required to switch from valley back to peak current. The positive pulse is called the turn-on pulse, the negative pulse the turn-off pulse. The td is used in a very simple circuit configuration to perform both the OR and AND logic functions. The only difference between the two circuits is that the tunnel diode OR gate input resistors are high, while the tunnel diode AND gate input resistors are of a low value.

a. The valley point of the td is the region where td current is at its _____ value.
b. Tunnel diode current will switch from its valley-point to its peak-point value when a (an) _____ pulse is applied.
c. The characteristic of the td where an increase in current through the td produces a decrease in the voltage drop across it is called _____ _____.
10. The PNPN device is constructed of four layers of semiconductor material, two N types and two P types. The operation of the PNPN device can be easily understood if it is compared to an NPN and a PNP transistor interconnected. The emitter-base junctions of both the NPN and PNP transistors are forward-biased. The collector-base junctions are reverse-biased. Fig. 5-31 shows that the two outside P-N junctions of the PNPN device are forward-biased and the center P-N junction is reverse-biased. The voltage-current characteristic curve of the PNPN device shown in Fig. 5-32 shows that when the forward-bias voltage reaches the breakover value, PNPN forward current will switch from cutoff to saturation. Once the device is conducting, forward-bias voltage can be reduced to the latching value and there still will be saturation current. Only when the forward-bias voltage is reduced below the latching value will the PNPN device switch from saturation to cutoff. The major application of the PNPN device is the silicon controlled switch (scs). The scs can consist of a two-element anode-cathode device, or by providing external connections to the inner elements, either a three- or four-element device. The inner element connections are called the cathode gate and the anode gate. The scs is a low-voltage, low-current switching device compared to silicon controlled rectifier (scr) which is a high-voltage, high-current switching device.
a. The four external connections of the scs are the _____, _____ _____, _____, and _____ _____.
b. The minimum value of voltage required to maintain saturation current of a PNPN device is called the _____ voltage.

SUMMARY ANSWERS

1a. The quantity of impurities added to a zener diode determines the critical value of applied **reverse** bias to which the diode reacts.

1b. Reverse current through a zener diode is high when the proper **breakdown** voltage is applied.

1c. When reverse current through a zener diode is high, the forward current is **zero**.

2a. If the breakdown voltage of the diodes shown in Fig. 5-7 were not equal, the **amplitude** of the output-voltage waveform would be affected.

2b. If the voltage applied to the circuit shown in Fig. 5-8 were reversed, high **forward** current through the zener diode would divert current from the meter circuit.

3a. The direction of current when the switch shown in Fig. 5-9 is opened is **opposite to** the current supplied by the battery.

3b. If the input voltage shown in Fig. 5-10 is reversed the voltage across the load would be approximately **zero**.

4a. The capacitance of a varactor is dependent on the magnitude of **reverse-bias** voltage.

4b. When the depletion area of a varactor is increased, the capacitance is **decreased**.

4c. When the intensity of light focused on a photoresistor is increased, the current through the associated circuit is **increased**.

5a. If the source of light is kept from reaching the P-N junction shown in Fig. 5-14, the relay contacts will **close**.

5b. With switch S1 closed, the photodiode shown in Fig. 5-15 can be excluded as a source of trouble by shorting the photodiode leads and observing that the high-beam lights are switched **off**.

5c. Four solar cells, of the type described, when connected in series-parallel will supply 0.8 volt, providing a source of light is available.

6a. The resistance of the scr is **high** until a gate pulse is applied.

6b. The scr is used as a switching device in **high-voltage** circuits.

6c. With a gate pulse applied, the cathode must be **negative** to cause the scr to conduct.
7a. The three elements of the ujt are **base 1, base 2,** and **emitter.**
7b. When a positive pulse voltage is applied between emitter and base 1 of the ujt, the emitter-base 1 junction will become **forward**-biased.
7c. The operational region of peak-point current will be reached only when the emitter-to-base 1 junction is **forward**-biased.
8a. The time constant of the resistor and capacitor in the ujt emitter-base 1 circuit determines sawtooth-voltage **frequency.**
8b. The emitter-to-base 1 resistance of the ujt is low when base 1-to-emitter current is at its **peak** value.
8c. The emitter diode of a ujt multivibrator is reverse-biased while the ujt is **conducting.**
9a. The valley point of the tunnel diode is the region where td current is at its **minimum** value.
9b. Tunnel diode current will switch from its valley-point to its peak-point value when a **turn-off** pulse is applied.
9c. The characteristic of the td where an increase in current through the td produces a decrease in the voltage drop across it is called **negative resistance.**
10a. The four external connections of the scs are the **anode, anode gate, cathode** and **cathode gate.**
10b. The minimum value of voltage required to maintain saturation current of a PNPN device is called the **latching** voltage.

FINAL TEST

1. The ten arabic symbols (0 through 9) are used to represent quantities in the _____ number system.
   a. binary
   b. octal
   c. decimal
   d. Roman

2. The digits 17 in the octal number system represent the quantity _____ in the decimal number system.
   a. 8
   b. 14
   c. 15
   d. 17

3. Since the base or radix of the binary number system is two, only _____.
   a. the digit 2 is used
   b. the digits 1 and 2 are used
   c. the digits 0, 1, and 2 are used
   d. the digits 0 and 1 are used

4. The digit 3 in the decimal number 345 represents the quantity _____ in the decimal system.
   a. 345
   b. 300
   c. 30
   d. 3

5. Binary 1111 represents the quantity _____ in the decimal number system.
   a. 1111
   b. 16
   c. 17
   d. 15

6. The digit 1 in binary 1000 represents the quantity _____ in the decimal number system.
   a. 8
   b. 4
   c. 2
   d. 1

7. The digit 1 in binary 0100 represents the quantity _____ in the decimal number system.
   a. 8
   b. 4
   c. 2
   d. 1

8. Decimal 28 equals binary ___.
   a. 00111
   b. 00028
   c. 11101
   d. 11100

9. If the least significant digit of a binary or decimal number is 0, the total quantity expressed must _____.
   a. be 0
   b. be 1
   c. be odd (not divisible by decimal 2)
   d. be even (divisible by decimal 2)

10. Binary 100000 equals decimal _____.
    a. 50,000
    b. 64
    c. 32
    d. 16

11. When a positive voltage is applied to the anode, and a negative voltage to the cathode of a semiconductor diode, it is _____-biased.
    a. zero
    b. negative
    c. reverse
    d. forward

12. The one characteristic that is common to all three of the basic transistor circuit configurations is _____.
    a. voltage gain

b. input impedance
c. power gain
d. phase relationship

13. The transistor logic circuit that has two or more inputs, a single output, and provides an output when any input signal is present is the _____ gate.
    a. AND
    b. OR
    c. NOR
    d. NAND

14. If the output of a three-input emitter-follower AND gate is +5 volts, input A is _____, input B is _____, and input C is _____.
    a. −5V, +5V, +5V
    b. +5V, +5V, +5V
    c. +5V, −5V, −5V
    d. −5V, +5V, −5V

15. The transistor logic circuit that has two or more inputs, a single output, and provides a signal output only when all input signals are present, is the _____ gate.
    a. AND
    b. OR
    c. NOR
    d. NAND

16. The NOT circuit will _____ its input signal.
    a. distort
    b. inhibit
    c. amplify
    d. invert

17. When all transistors of an NPN direct-coupled NAND gate are conducting the output will be _____.
    a. zero volts
    b. same as the input
    c. a logical 1
    d. a logical 0

18. The exclusive OR circuit functionally consists of the _____ circuits.
    a. AND, NAND, and OR
    b. OR, NOT, and AND
    c. OR, AND, and NOR
    d. NOR, AND, and OR

19. The inhibitor circuit provides an output signal only when _____ are present.
    a. all signals except the inhibiting signal
    b. all signals
    c. the inhibiting signal and a positive signal
    d. the inhibiting signal and a negative signal

20. When input A is zero and input B is a minus voltage in a direct-coupled exclusive OR circuit, Q2 is _____ and Q3 is _____.
    a. conducting, cut off
    b. cut off, conducting
    c. conducting, conducting
    d. cut off, cut off

21. An oscillator circuit must include _____.
    a. amplification or positive feedback
    b. amplification and positive feedback
    c. an NPN transistor
    d. PNP transistor

22. When the time constant of a circuit is increased, the _____.
    a. frequency is increased
    b. circuit is not affected
    c. circuit becomes inoperative
    d. frequency is decreased

23. A one-shot multivibrator remains in the logical zero state until _____.
    a. operating power is applied
    b. external excitation is removed

199

c. an input pulse is applied
d. internal feedback takes over

24. A one-shot is held in the logical one state by _____.
   a. an external logical 0 pulse
   b. the internal RC network
   c. the external RC network
   d. an external logical 1 pulse

25. A more common name for the bistable multivibrator is _____.
   a. the Eccles-Jordan circuit
   b. a one-shot
   c. the flip-flop
   d. the d-c transistor circuit

26. A binary storage device is in the "1" state when _____.
   a. the reset input signal is present
   b. the device has been set
   c. the device has been reset
   d. the trigger input signal is present

27. When the "1" output of a flip-flop is high, the _____.
   a. 0 output is low
   b. 0 output is high
   c. flip-flop is reset
   d. trigger input signal is present

28. To change the state of an RS-type flip-flop from the reset state to the set state, a _____ _____ is required.
   a. complementary input
   b. trigger input
   c. reset input
   d. set input

29. When a flip-flop has stored a logical "0," it is _____.
   a. reset
   b. set
   c. ready to change state
   d. affected by a reset input

30. An asynchronous switching system does not include _____.
   a. flip-flop
   b. logic circuits
   c. binary storage devices
   d. a clock

31. If the amplitude of the input signal shown in Fig. 4-6 is increased, the _____.
   a. output frequency changes
   b. output amplitude increases
   c. pulse repetition time changes
   d. pulse width is increased

32. The output of a differentiator circuit is across _____.
   a. the series-connected RC network
   b. the capacitor
   c. the resistor
   d. the input

33. The output of an integrator circuit is across _____.
   a. the series-connected RC network
   b. the capacitor
   c. the resistor
   d. the input

34. When flip-flops 1 and 3 are set (Fig. 4-9) and flip-flops 2 and 4 are reset, the circuit has received _____ pulses.
   a. 9
   b. 3
   c. 7
   d. 5

35. When flip-flop E is set (Fig. 4-11 and 4-12), the state of all the ring-counter stages can be represented by _____.
   a. 01000
   b. 10000
   c. 11000
   d. 10001

36. A ring counter that has the capacity to count from 0

9 must have _____ flip-flop stages.
a. 5
b. 9
c. 10
d. 18

37. If the states of flip-flops D, C, B, and A of a decade counter are respectively, 0111, then the next states will be _____.
a. 1000
b. 1001
c. 1010
d. 0101

38. If an electronic counter indicates that 100 pulses are present during a 5-millisecond gating time, the input frequency under measurement is _____.
a. 500 Hz
b. 1000 Hz
c. 200 kHz
d. 1 MHz

39. A square wave with a pulse repetition time (prt) of .0008 second has a pulse repetition rate (prr) of _____ pulses per second.
a. 125
b. 250
c. 1250
d. 2500

40. If 32 negative pulses are applied to a divide-by-four parallel counter, _____ pulses will appear at the output of the circuit.
a. 2
b. 8
c. 16
d. 32

41. Electron multiplication by collision is a condition found in a _____ diode.
a. double-base
b. photo
c. tunnel
d. zener

42. When breakdown voltage is applied to a zener diode, reverse current is _____, and the voltage drop across the zener diode is _____.
a. high, constant
b. low, zero
c. zero, constant
d. high, zero

43. The three-element semiconductor device that consists of an emitter, base 1 and base 2, is the _____.
a. silicon controlled rectifier
b. unijunction transistor
c. tunnel diode
d. silicon controlled switch

44. The gated silicon controlled switch is used primarily in _____ _____, _____ control circuits.
a. high voltage, low current
b. low voltage, low current
c. high voltage, high current
d. low voltage, high current

45. When a high-intensity light source is focused on a photoresistor it will have a _____ resistance and _____ current through it.
a. high, high
b. high, low
c. low, high
d. low, low

46. The value of voltage applied to a PNPN device to cause it to switch from cutoff to saturation current is called the _____ voltage.
a. latching
b. breakdown

c. trigger
d. breakover

47. The increase in the capacitance of a varactor is accomplished by _____ voltage.
    a. lowering the reverse bias
    b. raising the forward bias
    c. raising reverse bias
    d. lowering the forward bias

48. If four solar cells are connected in series-parallel the output voltage will be _____ when a light is applied.
    a. .4V
    b. .8V
    c. 1.2V
    d. 1.6V

49. If input resistor A of a tunnel diode AND gate is open, tunnel diode anode current will _____.
    a. decrease to valley point
    b. increase to peak point
    c. decrease to zero
    d. remain at maximum level

50. If the breakdown of each of the two zener diodes connected back-to-back across the primary of a transformer is 60 volts, and the applied voltage is 130 a-c volts, the voltage across the primary is _____.
    a. 30V
    b. 60V
    c. 120V
    d. 130V

ANSWERS TO FINAL TEST

1.	c	18.	b	35.	b
2.	c	19.	a	36.	c
3.	d	20.	c	37.	a
4.	b	21.	b	38.	c
5.	d	22.	d	39.	c
6.	a	23.	c	40.	b
7.	b	24.	b	41.	d
8.	d	25.	c	42.	a
9.	d	26.	b	43.	b
10.	c	27.	a	44.	b
11.	d	28.	d	45.	c
12.	c	29.	a	46.	d
13.	b	30.	d	47.	a
14.	b	31.	d	48.	b
15.	a	32.	c	49.	d
16.	d	33.	b	50.	c
17.	d	34.	d		

Index

Addition
  binary, 32-33
  logical, 52
Addresses, 20
Amplitude-limiting circuits, 138-139
Analysis
  of timing diagram, 142
  of truth table, 93
AND gate, 46
  direct-coupled, 77
  emitter-follower, 76
  logic and equations, 90
  negative logic, 88
  positive logic, 88
Applications
  for sequential circuits, 127
  of counters, 154-155
  of solid-state logic circuits, 18-19
  of zener diode, 168-171
Arithmetic, digital, 16-17
Arithmetic element, 21
Automation, 18-19

Base or radix, 17, 25
Basic diode switch, 70
Basic elements of digital computers, 20-21
Basic flip-flops, 42
  symbols, 43
Basic symbolic logic operations, 50-53
  addition, 52
  combining logic circuits, 53
  multiplication, 50-51
  negation, 51
Biasing the common-base circuit, 74
Biasing the inverter circuit, 75
Biasing transistor, 74-75
Binary, 124
  addition, 32-33
  -coded decimal numbers, 38-39
  counters, 45

Binary—Cont'd
  numbers, representation of, 40-41
  subtraction, 34-35
  system, 23, 26-27
    base or radix, 26
    positional notation, 26-27
    -to-decimal conversion, 30-31
Bits, derivation of, 23
Breakdown voltage, 167

Cascaded binary circuits, 142-143
Change of state tabulation, 143
Characteristics, pulse, 137
Circuit
  amplitude-limiting, 138-139
  cascaded binary, 142-143
  combinational, 53, 54-55, 86-87
  differentiator, 140
  diode limiter, 138
  exclusive OR, 55
  inhibitor, 86
  integrator, 141
  logic, 46-47
  NOT, 47
  NOT AND, 54-55
  NOT OR, 54
  sequential, 53
  squaring, 139
Closed diode switch, 70
Code
  excess-3, 39
  natural binary decimal, 38
Collected-coupled multivibrator, 106-108
  circuit operation, 108-111
  waveform analysis, 112-113
Combinational circuits, 53, 54-55, 86-87
Combining logic circuits, 53
Common-base circuit biasing, 74
Complements, 36-37
Computer, stored program, 20
Constructing a truth table, 92-93
Control element, 21

Conversion
  binary-to-decimal, 30-31
  decimal-to-binary, 28-29
Counter
  applications, 154-155
  binary, 45
  decade, 45
  divide-by-five, 150-151
  divide-by-four, 146-147
  divide-by-three, 148-149
  parallel decade, 152-153
  preset, 44
  ring, 45
Counting
  direct, 154
  indirect, 155
Current versus electron flow, 68

Data words, 20
Decade counters, 45
Decimal
  code, natural binary, 38
  numbers, binary-coded, 38-39
  system, 24
  -to-binary conversion, 28-29
  -to-binary equivalents, 27
Delay time, 42
Derivation of bits, 23
Development
  of digital devices, 14-15
  of rectangular waveform, 104-105
Description of rectangular waveform, 104
Differentiator circuit, 140
Digit, 14
Digital
  arithmetic, 16-17
  computers, basic elements of, 20-21
  devices, development of, 14-15
Diode
  AND gate, 71
  limiter circuit, 138
  logic, 70-71
  OR gate, 71
  switch, basic, 70
  zener, 166-167
Direct
  counting, 154

Direct—Cont'd
  -coupled AND gate, 77
  -coupled exclusive OR circuits, 84-85
  -coupled NAND gate, 80
  -coupled NOR gate, 82
  -coupled OR gate, 79
Divide
  -by-five parallel counter, 150-151
  -by-four logic, 146-147
  -by-four parallel counter, 146-147
  -by-three parallel counter, 148-149
  -by-three timing diagram, 149
  -by-two method, 29
Double-dadd method, 31
Duality of logic circuits, 88-89

Eccles-Jordan circuit operation, 122-123
  application of first trigger pulse, 122
  application of second trigger pulse, 123
Eccles-Jordan multivibrator, 120-121
  circuit configuration, 121
Element
  arithmetic, 21
  control, 21
  input, 21
  output, 21
  storage or memory, 21
Emitter-follower
  AND gate, 76
  OR gate, 78-79
Excess-3 code, 39
Exclusive OR
  circuit, 55
  logic levels, 85
  voltage, 85

Flip-flops, 42
  logical, 124-125
  RS-type, 125
  symbols, 43
Forward bias, 69

Gate
    AND, 46
    diode AND, 71
    diode OR, 71
    OR, 46
Generation of pulses, 136-137

Indirect counting, 155
Inhibitor circuit, 86
    truth table for, 87
Input element, 21
Instruction words, 20
Integrator circuit, 141
Inverter circuit, biasing of, 75

Latch, 124
Logic
    diode, 70-71
    divide-by-four, 146-147
    negative, 81
    positive, 81
    symbolic, 48-53
Logical
    addition, 52
    flip-flop, 124-125
        basic specifications, 124
        RS-type, 125
    multiplication, 50-51
Logic and equations
    AND gate, 90
    NAND gate, 91
Logic circuits, 46-47
    and symbolic logic, 48-49
    duality of, 88-89

MAP, 48
Materials
    N-type, 68
    P-type, 68
    semiconductor, 68
Method
    divide-by-two, 29
    double-dadd, 31
Memory, 127
Monostable multivibrators, 43
Multiplication, logical, 50-51
Multivibrator
    Eccles-Jordan, 120-121
    monostable, 43
    one-shot, 42, 114-117
        symbols, 43
    why required, 104-105

NAND gate
    direct-coupled, 80
    logic and equations, 91
Natural binary decimal code, 38
Need for binary number system, 17
Need for symbolic logic, 48
Negation, 53
Negative logic, 81
    AND gate, 88
NOR gate
    direct-coupled, 82
    logic levels, 83
    voltage, 83
NOT AND circuit, 54-55
NOT circuit, 47
NOT OR circuit, 54
Notation, positional, 26-27
N-type materials, 68
Numbers, binary-coded decimal, 38-39
Number systems, 16, 22-27

One-shot multivibrator, 42, 114-117
    application of trigger pulse, 115
    circuit variations, 117
    frequency and output pulse duration, 116
    initial-state operation, 114
    return to initial state, 116
    symbols, 43
    waveform analysis, 118-119
Open diode switch, 170
OR circuits, direct-coupled exclusive, 84-85
OR gate, 46
    direct-coupled, 79
    emitter-follower, 78-79
    positive logic, 89
Oscillator, sine-wave, 136
Output element, 21

Parallel decade counter, 152-153
Photodiode, 174
Photoresistor, 173
Photovoltaic cells, 175
Pip, 140
PNPN devices, 186-189
    characteristic curve, 188

PNPN devices—Cont'd
  operation of, 187
  symbolic representation, 186-187
Positional notation, 26-27
  system, 24-25
Positive logic, 81
  AND gate, 88
  OR gate, 89
Preset counter, 44
Programmer, 20
P-type materials, 68
Pulse
  characteristics, 137
  generation, 136-137
  waveforms, 136

Real-time control, 19
Rectangular waveform, 104-105
Rectifier, silicon controlled, 176-177
Representation of binary numbers, 40-41
Reverse bias, 69
Ring counter, 45, 144-145
RS-type flip-flop, 125

Schmitt trigger, 42
  symbols, 43
Semiconductor
  diodes and biasing, 68-70
  materials, 68
  theory, 166
Sequential circuit, 53
  applications, 127
  types, 126
Silicon controlled rectifier, 176-177
  applications of, 177
Silicon controlled switch, 189
Sine-wave oscillator, 136
Solid-state logic circuit applications, 18-19
Special purpose semiconductors, 172-175
Special semiconductor circuits, 56-57
Spike, 140

Squaring circuit, 139
Storage or memory element, 21
Stored program computer, 20
Subtraction
  binary, 34-35
    complement method, 37
  with true complement method, 36-37
  with radix-minus-one complement, 37
Switching time of tunnel diode, 184
Symbolic logic, 48-53
  need for, 48
Symbols, transistor, 72
System
  decimal, 24
  number, 16
  positional notation, 24-25
  tally, 22

Tabulation, change of state, 143
Tally system, 22
Thyristor, 57
Timing diagram analysis, 142
Toggle, 124
Transistor
  biasing, 74-75
  symbols, 72
  types and configurations, 72-73
Trigger pulse, 140
Truth table
  analysis of, 93
  construction of, 92-93
Truth tables and logic, 92-93
Tunnel diode, 182-183
  AND gate, 185
  OR gate, 184
  switching time, 184

Unijunction transistor, 178-181
  circuit applications, 180-181

Varactor, 172-173

Waveforms, pulse, 136

Zener diode, 166-171